Interpreting *Environments*

Tradition, Deconstruction, Hermeneutics

T

onstructio

Interpreting
Environments

Tradition,
Deconstruction,
Hermeneutics

Robert Mugerauer

 University of Texas Press, Austin

Library of Congress Cataloging-in-Publication Data

Mugerauer, Robert.
    Interpreting environments : traditions, deconstruction, hermeneutics / Robert Mugerauer.
        p.   cm.
    Includes bibliographical references and index.
    ISBN 0-292-75178-8 (permanent paper). — ISBN 0-292-75189-3 (pbk.)
    1. Human ecology—Philosophy. 2. Landscape assessment—Methodology. 3. Human geography—Philosophy. 4. Human geography—Methodology. I. Title.
GF21.M83 1996
304.2—dc20   95-8156

Dedication

Dedication

To Louis Mackey for his exegetical exercises.

To Richard Zaner for showing how the professions hold rich clues to understanding the embodied self in the world.

To J. B. Jackson for sharing his sense of wonder at the vernacular.

Contents

Contents

Illustrations

Illustrations

Acknowledgments

Acknowledg

By supporting the development of theory in the curriculum and research agenda of the School of Architecture at the University of Texas at Austin, Dean Hal Box helped to make this work possible. I want to thank also the following faculty members and scholars who provided honest criticism, valuable suggestions, and encouragement during the many rewritings of these essays: Michael Benedikt, Robin Doughty, Ken Foote, Roderick Lawrence, David Saile, David Seamon, Anne Vernez-Moudon, Andy Vernooy, and Fran Violich.

Dr. J. Gray Sweeney, a colleague for over twenty years, provided a great amount of substantive and collaborative material on nineteenth-century American landscape painting from an entirely independent project he was pursuing as senior fellow at the Smithsonian Institution's National Museum of American Art, continuing our satisfying habit of working together on mutual interests. Not surprisingly, in the course of writing chapter 3, "Hermeneutic Retrieval: American Nature as Paradise," particularly the parts of sections 2 and 3 that provide close explications of landscape paintings, I often wound up following his interpretations closely and occasionally used wording from his out-of-print *Themes in American Painting*. Because he graciously proposed that I forgo the usual methods of attribution in these cases, arguing that a jumble of quotation marks would be confusing and unnecessary, and because publishing conventions do not easily allow a way to indicate a collaboration in a small section of a larger and diverse work, he merits special recognition as a contributor to these sections, though he is not in any way responsible for their shortcomings. Material from *Themes in American Painting* is used with full permission of Professor Sweeney, its copyright holder.

Richard Etlin is to be acknowledged for his cheerful attitude to my characterization of his interpretation of French neoclassic pyramids as an "establishment" position and to my subsequent deconstruction of that interpretation in the course of chapter 2.

I appreciate the following colleagues, who helped me to obtain illustrations: Rajesh Gulati, Kevin Keim, Martha Leipziger-Pearce, J. Gray Sweeney, Dana Norman, and Andy Vernooy.

Finally, I am delighted to express appreciation for the thoughtful help of my colleagues at the University of Texas Press.

Interpreting *Environments*

Tradition, Deconstruction, Hermeneutics

Introduction

Although we too often take for granted the built and natural environments in which we live, we do attend to them when they aggravate or please, and at times we become fascinated with them. When going about our business and enjoying our leisure, and especially when traveling, we encounter buildings and landscapes from other times and places and wonder, "How could they live like that?" or perhaps, "Why don't we do things that way?" In our civic lives we debate whether neighborhoods, public places, cities, and whole regions should change, and if so, how. Apparently, people disagree not only about how things should be but even about how they are (or were).

Some of us have found these issues so engaging that we have chosen careers in environmental disciplines and professions, but nearly everyone spends some time trying to figure out why environments are the way they are, how they might be otherwise, and what difference it would make if they were. Such questions are both inherently interesting and practically critical, because we are intrigued by the world's variety and because historical environments manifest our basic hopes and fears. Answering these questions requires us not only to articulate what we desire to achieve for ourselves and to leave for others but also to assess the cultural and physical barriers to these goals. Thus, the environment stimulates imagination and critical analysis as we ponder our own and others' daily routines, extraordinary experiences, and past and future ways of living—that is, alternative visions of the cosmos and orientations in the world.

Just beneath the surface of these issues are complex, contested theoretical questions about what sorts of meaning environments might have and practical questions about how we might discover and apply such meanings. Today these problems are especially pressing, even confusing, because we are in the midst of pluralistic, skeptical, and radical challenges to the assumptions and approaches traditionally used for environmental interpretation.

The scope of these challenges, then, provides the primary motive for this set of essays: as a philosopher who has migrated to teaching and research in architecture, planning, geography, and American studies, I am responding to colleagues and students from a variety of environmental disciplines who want to understand the fashionable—almost mandatory—Continental methodologies such as deconstruction and hermeneutics. Although keenly interested, many such individuals simultaneously are frustrated by these approaches' neglect of the physical environment and tired of trying to adapt the strategies and vocabulary of literary interpretation for use in discussing environmental issues. Researchers at all levels of accomplishment regularly ask, "Why and how would deconstruction be important to my work?"; "Hermeneutics seems so vague and difficult; how would I use it?"; or "How can I decide whether these approaches are preferable to traditional methodologies?" It is frustratingly hard for such individuals to master the theory and apply it to the environment without clear, sustained examples in their own areas of interest and expertise. This book is meant to remedy that situation by providing a kind of handbook of traditional, deconstructive, and hermeneutic interpretation.

My goal is to show how traditional, deconstructive, and hermeneutic approaches go about interpreting the environment, not to compare, evaluate, or judge the alternatives or to persuade readers of the merits and deficiencies of each. Thus, in each essay the position that has the floor speaks with its own voice and intention. There is no cumulative argument, no overall thesis to be proven. My own view is absent—as it is for many a teacher. Readers can make up their own minds about whether or how to proceed with these exceedingly dense approaches.

Some readers—those who are already familiar with the approaches or interested not in more theory but only in any differences that can be discerned in practicing interpretation—can skip most of this introduction. Perhaps a quick glance at the final section, which provides a guide to the book's organization, might be useful. The reader can then proceed to whatever chapter seems most interesting or relevant.

Other readers may prefer to have a kind of primer of the three contending approaches. Certainly, a sense of the historical situation, of the main areas of agreement and disagreement among the tradition, deconstruction, and hermeneutics, will make the import of the different interpretations clearer. Because in the three interpretive chapters I do not pause to describe or explain the theories being applied, readers wanting to understand the three approaches' main features, their beliefs, strategies, and points of contention, can read through the introduction and then move on either through the book or to whatever chapter seems to be the most appropriate.

The Historical Context

The inaccessibility of contemporary theory that environmental researchers and professionals experience is due in large part to the interpretative methodologies' diffusion patterns and overwhelmingly linguistic emphasis. In the 1970s philosophers and philosophically trained theorists began a major shift away from traditional approaches and formalism. This revolutionary work by Martin Heidegger, Hans-Georg Gadamer, Michel Foucault, Jacques Derrida, and others soon spread to the closely related fields of literary theory and criticism and comparative literature because it privileged language and provided strategies for reading that offered an almost entirely new way of making sense out of texts—a desirable prospect to experienced scholars who wanted a fresh way to teach and to all who appreciated that these new approaches would let them produce new readings of canonical texts and would require a new cohort of academic specialists. The movement further spread to other disciplines in the humanities and social sciences concerned with "writing culture," such as history, sociology, and anthropology, and then to area studies and, less successfully, art history.[1]

During the 1980s a second wave of work sought to build on or displace the first. *Postmodernism* and *poststructuralism* arrived as at least vaguely understood descriptors. Jean-François Lyotard, Jean Baudrillard, Gilles Deleuze and Felix Guattari, Michel de Certeau, Luce Irigaray, and Julia Kristeva developed the historical, psychological, economic, social, political, and gender dimensions of processes and practices in the postmodern post-subject/object era. In response, Jürgen Habermas and Alasdair MacIntyre argued, respectively, on behalf of the modern and classical traditions.[2]

An unexpected and energizing war of ideas was under way, but it did not spread across all academic disciplines and professional practices uniformly. Only after the linguistically based disciplines had substantially shifted did architecture, urban planning, environmental design, landscape studies, and cultural geography gradually begin to push beyond the dominant emphasis on language and textuality and pick up on the relatively obscure redefinitions of things, space, and the built environment contained in the new approaches. Even then, however, the newer methods of the 1970s and 1980s were not immediately used, because more familiar varieties of Marxism and phenomenology already were being used to explicate space and built environments. Naturally, in the hothouse that was nurturing theory, the already planted approaches blossomed quickly and were the first to bear fruit for environmental research.

The Marxist-inspired work included Walter Benjamin's analyses of urban life, cities, and streets that became easily available to English speakers with the publication of *Reflections* in 1978; his famous, fragmentary Arcades Project was never finally finished but was published as a "reconstruction" in 1989. Henri Lefebvre's *Everyday Life in the Modern World* was translated in 1971; that out-of-print work has just been reissued and his *Production of Space* made available in English. Fredric Jameson has especially influenced architects and planners since the late 1980s; David Harvey and Edward Soja both published major works in 1989 that develop post-structuralist interpretations of culturally constituted space; Denis Cosgrove and his colleagues produced several books at the end of the 1980s. Thus, although Marxist-connected approaches to space, buildings, and landscapes have been present all along, the major impact of these cultural critiques and analyses of postmodernism is just now broadly rippling through the environmental and spatial disciplines.[3]

Strong advances also were made in the phenomenology of place and environment. Christian Norberg-Schulz was among the first to open a sphere for Heidegger's influence in architecture and landscape in the late 1970s and early 1980s. Karsten Harries, a philosopher at Yale teaching and writing about architecture, also made an early, seminal impact and consolidated phenomenology's importance for architecture. In geography and behavioral-environmental research, Anne Buttimer, Edward Relph, and David Seamon have made valuable contributions and reached a wide audience for over a decade.[4] Paradoxically perhaps, their very success in adapting phenomenology to the study of place, environment, and dwelling led researchers to remain with phenomenology and not go on to examine the more radical approaches of hermeneutics or deconstruction.

Today, even as the phenomenological and Marxist-based interpretations become accessible and the newer modes of thought spread, the difficulty of hermeneutics, deconstruction, and poststructuralist approaches remains a fundamental problem for environmental professionals. Heidegger's thought, for example, is notoriously opaque and nonlinear. Even many professional philosophers say he makes no sense. Heidegger's followers, such as Gadamer, are only a bit easier. Movements such as Derrida's deconstruction seem even more arbitrary or centripetal and disturbingly skeptical, if not cynical. And with the very latest and exotic work appearing as quickly as the publication industry can put it in print, it seems impossible to catch up, much less keep up.

For all this, at least the ideas are available: the primary works are being translated and published and reliable secondary sources are beginning to appear—some of the latter even written by environmental professionals, although it is often obvious that they have their theory at second hand.

To present thorough interpretations, I limit myself to providing a heuristic guide to the three major approaches—the tradition, deconstruction, and hermeneutics—because giving short examples of all the proliferating methodological varieties would only replicate the current, confusing situation. The simplification is further warranted because these three powerful approaches provide the fundamental alternatives that the proliferating exotics elaborate.[5]

Even if the methodologies themselves are increasingly intelligible, however, it still is not clear how they apply to the built environment rather than to language, texts, and psychological-sociological practices. Although scholars are starting to work out the possibilities of these methodologies, especially in graduate-level university research and avant-garde journals such as *Assemblage* and *Threshold,* the available analysis of buildings and landscapes is fragmentary, scattered, and sometimes superficial or untrustworthy. All the activity has produced only a few sustained analyses that join theoretical mastery to professional familiarity with the built environment, especially with regard to non-avant-garde work.[6] Remarkably, sustained examples applying the two most powerful recent approaches, hermeneutics and deconstruction, to the environment are not available.

When asked how theory would change environmental interpretation, criticism, and practice, practitioners of hermeneutics and deconstruction can point to only a handful of primary examples: Heidegger's overcited passages on the Greek temple in "The Origin of the Work of Art" and on the Black Forest farm buildings and the fourfold of earth, heavens, mortals, and divinities in "Building Dwelling Thinking" and Derrida's less well known comments on pyramids, his work with Peter Eisenman on a folly for Bernard Tschumi's Parc de la Villette in Paris, or his comments on unbuilt deconstructive architectural designs.[7]

Deconstruction, although not hermeneutics, additionally repels environmental researchers with its view that texts and practices are finally infra-referential cultural artifacts without any ultimate extratextual reference to an assumed "external reality." Such a hermetic vision, even if it embraces the play of all historically significant signifiers, does not obviously apply to buildings and mountainsides, which appear to be a "primary external reality," secondarily re-presented by poems, diaries, paintings, and so on. Thus, insofar as hermeneutics and deconstruction both focus on "texts" and the latter argues something close to the position that all the world is "language-signifier," the environmental disciplines are not immediately engaged.

The Three Alternatives

There are historical and principled reasons for these three approaches having developed as they have and commanding the stage of today's debates about what texts and environments mean, what interpretive work is good, and who will succeed professionally. Both the history and the "logic" of the principled differences order the three approaches in particular ways, although I deliberately follow none of these in arranging the chapters, to avoid tacitly agreeing with any one of the "metathinkings."

The Tradition

Traditional Western interpretation of the arts and the built environment has remained vital because it focuses on two basic relationships that humanly produced works have to their natural and cultural contexts. As described in time-tested metaphors, what we make and interpret is both a "mirror" and a "lamp," because it reflects the reality from which it derives and creatively illuminates that reality.[8]

Throughout the variations of twenty-five centuries of explicit aesthetic and critical theory, the basic foundation remains the same: what we make has meaning because of its extrinsic relations. Plato analyzed and described metaphysical, epistemological, and ethical relationships in terms of hierarchy and participation. He argued that what is humanly fashioned re-presents ideal forms; our ability to discern the differences between the timeless ideals and changing temporal and spatial appearances initiates a circle, or upward spiral, of understanding, enabling us to come closer to excellence.

Aristotle put the same idea in terms of principles of intelligibility, or causes. He held that we could understand what we make (*technē* and *poesis*) in terms of the efficient source or human agent responsible for it, its form, its materials, and its final goal or function. Both theories involved the framework that has been developed since: understanding a work depends on interpreting it in the light of its origin or creation, its forms, materials, and content, and its ethical and intellectual impulse back to social, natural, and perhaps sacred reality.

In the Enlightenment the concept of representation was transformed, with differing emphases, in the United Kingdom and on the Continent. David Hume and Edmund Burke shifted attention to an anthropological or psychological correlation between made objects and our private and social experiences. Hume spoke of meanings in terms of our sensory perceptions and our idiosyncratic and shared customs of association and judgment. Burke analyzed environmental responses to what we call the beautiful and sublime in terms of fundamental emotions such as fear, pleasure, and love.

Immanuel Kant, on the other hand, in analyzing the relation of consciousness to sensory data in generating coherent experiences, worked out the manner in which environmental objects have meaning by representing previously made cultural forms and types. Architecture, he argued, represents not naturally occurring forms but the humanly invented, which has no precedent in nature—doors, arches, temples, and so on. G. W. F. Hegel moved back to the metaphysical tradition in arguing that meaning is generated precisely by absolute Mind (ultimate reality) historically manifesting itself in and through cultural products. Architecture, broadly understood, manifests the phases of the unity of spiritual meaning and material forms in such a way that the epochal changes of what we build provide the means for us to become conscious of the historical unfolding and progress of the universe. On the basis of this correspondence, J. J. Winckelmann worked out a more detailed idea of artistic styles as reflecting historical change and thus being the basis for our current theories of art-historical periods.

Even without an elaboration of these issues, it is clear why such sophisticated interpretation depends on objective correctness in regard to the form and content, or "material-symbolic" dimensions, of the work in question. Only when "preinterpretive"—usually technical—scholarship guarantees that we are certain about what a thing is (a previously lost original, a derivative copy, the result of cultural diffusion across space, etc.) can we proceed to interpret the relation between the work and its original and historically developed context. Not surprisingly, traditional interpretation cooperates with disciplines such as archaeology, philology, historiography, intellectual history, material and technology studies, and historical restoration and preservation.

Once scholars know what they are analyzing, the fundamental interpretive move is to discern and work out in detail the ways in which the work mirrors the dimensions that produced it. The meaning that interpretation seeks lies in the connections between the past or present external conditions of existence and the work's internal features or characteristics. Of course, the work may be seen as the product of any or many of the forces that generate things in the world. Normally this is very complicated; not only do people produce works, but they do so in response to a variety of factors: external forces of which they may or may not be aware, personal needs and desires, the material and formal features of the work itself as it emerges (either manifesting or resisting the original impetus or suggesting new possibilities that were not expected), and their own criticism and that of others.

Hence, the built environment can be seen either as the fairly anonymous product of the cultural forces and practices in effect at the time or as the result of the deliberate and creative effort of a particular creator or even "genius" (or of a small group of collaborators). As psychiatry and psychology have shown us, the creator may shape the work not only according to self-conscious intentions but also in ways of which she or he is unaware (because of the influence of the personal or collective unconscious). Alternatively, all these dimensions may play a part, in dizzying interaction.

To follow the idea that artificial objects mirror their contexts, insofar as impersonal economic, social, linguistic or symbolic, historical, and technological dynamics or personal factors shape what we do, the meaning of a work lies in the ways in which it represents its origin. We will be able to understand it insofar as we can follow its generation from its source (which we understand independently and objectively) and explain what we find in the work in terms of that origin.

Moving in the opposite direction along this same line of representational relation, we can also use an artifact to look back into its origin. That is, we can use what has been made to inform ourselves about the lost or obscured world from which it came. Historians, anthropologists, sociologists, and cultural and area studies analysts can learn about their subject matter (some dimension of the personal or cultural realms) by seeing how it was or is reflected in the works. Here, the works become repositories of meaning, waiting for the right interpretation to unlock what they have to reveal about the times, places, and conditions of their births and their subsequent histories.

In either case, it is interpreters' task and opportunity to travel the road of representation, working out and explicitly demonstrating how a work and its containing reality are related. It is the mutual dynamic between the external forces and internal features that is important and that enables us to use each one to understand the other.

Naturally, to be successful it is crucial to avoid merely speculating or generalizing about what might be plausible relations. Rather, it is necessary to find and display the objective correlation between the work and its objective circumstances. The actual, historical, and causal relation is sought. The interpreter needs to put aside personal prejudices and assumptions, to restrain supposition and unfounded inference in favor of explicating and proving the relations that obtain. Something very akin to the scientific method applies here. There has to be a sensitivity to the details of the phenomenon itself and a hypothetical explanatory schema that can be proved, or at least disproved, to account adequately for the features of the work (or of the contextual reality, if the reading is proceeding in the other direction). Essentially, in a way that parallels what science does for natural phenomena, interpretation provides a detailed, empirical exercise in discovering and demonstrating (often indirect and complex) causal relationships between the humanly made work and its contexts and origins.

This is possible not only for what were self-conscious factors during the creation of the work but also for unconscious or structural features that escaped the makers' attention or understanding. Insofar as unconscious concerns and limitations shaped the work, the causal connections need to be *and can be* proven by psychological or structural analyses of the influence of psychic life, religious and cultural beliefs, economic and historical forces, typological and material conventions, and so on, of which the creators may not have been focally aware. Thus, whether the forces that the work reflects are conscious or unconscious, personal or anonymous, the meaning is found when the interpretation demonstrates the connection, showing in concrete detail the fact and manner of the representation.

The other dimension—the work as illuminating the world—

involves not so much how the work tells us something about the reality that it necessarily mirrors as the ways in which the work is a source of original insights into reality. That is, in the classical and romantic traditions, the work is understood (respectively) as manifesting and creating new meaning. In a manner parallel to the way that a dramatic personal experience, friend or teacher, or discursive text can instruct us, the work shows us new dimensions of or possibilities in the world. We might learn a moral about human life or the nature of our existence. We might learn about what cannot be put into direct language or other symbolic form. For example, just as other people or our own experiences might help us to see and understand love or death, so too would the characteristics of the hearth and marriage bed or the cemetery and memorial do this.

Both intellectual and ethical interpretations of works explore the meaning of artifacts in terms of what we might learn from pondering or interpreting them ourselves. Thus, traditional interpretation is a more refined or systematic version of what it is to be hoped that we all do: examine the important works from the past and present to seek insight into ourselves, others, and the world about us. Works are deemed "classics" usually because they timelessly (or at least, for a long time and across cultures) shed light on the human condition and provide inspiration and consolation. Works are judged important when they provide deeper and more profound understanding than does most of what we hear, read, and see. Such artworks and environments are worth the interpretive effort because they pay us back by enlarging our minds and characters.

Methodologically, ethical or moral criticism interprets the insights and human vision available in built environments by making explicit what judgments or understandings may result from the work, by analyzing how the works provide this resource so that we may more fully and easily benefit from it, and by explicitly leading us to imagine and reflect on what "possible" worlds might unfold from the environment. Theorist and critic Yvor Winters argues that an artistic work "should offer a means of enriching one's awareness of human experience and of so rendering greater the possibility of intelligence in the course of future action; and it should offer likewise a means of inducing certain more or less constant habits of feeling, which should render greater the possibility of one's acting, in a future situation, in accordance with the findings of one's improved intelligence."[9] In interpretation we are asked to explore the work's meaning that might modify our lives by reflecting on the assumed and implied modes of life and courses of human action and by judging the conditions, responses, and responsibilities that might result.[10]

In the end, it is easy to see why the historical and biographical dimensions of meaning are so vital for the tradition. Because works are produced by one or more persons, it is necessary to make the connection between the works and their creators' successful or unsuccessful intentions, the features of the creators' individual or collective unconscious that are manifest without the creators' having been aware of them, and the autonomous historical or structural features (material, technological, economic, social, symbolic, political, etc.) that either supplement or overwhelm the creators' distinctive contributions. Similarly, insofar as the work illuminates life either for its makers or for its interpreters, its meaning lies in the connection between it and our experiences. Historical and biographical analysis is important in analyses along any of these dimensions. After all, the meaning of the work will lie not just in its internal formal configuration but in its relation to past and present human life.

Hermeneutics

Within the contemporary Continental tradition of hermeneutics—the theory and practice of interpretation—Martin Heidegger (whose shift beyond phenomenology to radical hermeneutics is only now becoming appreciated),[11] Hans-Georg Gadamer, Paul Ricoeur, and others attempt to provide an account of how the human sciences operate. Hermeneutics aims not so much to develop a new procedure as to clarify how understanding takes place.[12] It appears radical and has shaken traditional approaches mainly because it attempts to show the limitations and even groundlessness of what has been taken for granted. The project looks different from traditional scholarship because it focuses on what usually is taken as peripheral and critically brings to the foreground what usually is hidden or transformed in temporal divergences.

Hermeneutics points out the impossibility of scientific historiography's goal: to transcendentally and objectively pass over into another time to understand an earlier situation, text, or object in the same way that people of the time did. Such "objective knowledge would depend on a standpoint above history from which history itself can be looked upon," a position that finite humans cannot obtain.[13]

Still, hermeneutics also involves a belief that shared understanding is possible, both within and across traditions. Interpretation is a matter neither of finding the "one right interpretation," as the tradition contends, nor of calling attention to the interpreter's language and wit, where "everything is possible," as deconstruction holds, but of finding the valid criteria for polysemy within the fluid variety of possibilities. Beginning with our unavoidably finite and bounded situation, Gadamer develops the ways in which we can and do achieve nonarbitrary understanding: remaining open to the meaning of another person, text, and so on, to "what the other really is saying."[14]

According to hermeneutics, all understanding is interpretation, that is, contextual. Meaning always is produced in a specific time and culture, by finite humans. Because the context continually changes, no simple or fixed thing or meaning ever is there without interpretation, and we also have the obvious problem of our relation to other contexts.[15] This temporal, cultural context of our lives and meanings is called the "horizon" of understanding, because it is "the range of vision that includes everything that can be seen from a particular vantage point."[16]

To be human is constantly to attempt to understand, that is, to interpret things, to project expectations, and to discover whether and how those expectations are fulfilled. Because we always approach things and texts from within the horizons of what we are able to attend to, from within our time and place, with certain expectations about the existence and manner of their meanings, understanding naturally has presuppositions. We are prejudiced when we listen to another person talk or when we pick up a text; when reading a letter from home, for example, we might start by assuming that it will bring us news concerning something about which we care. We proceed from the preparation to hear and understand.

Heidegger made explicit the function of our expectations and assumptions by developing the idea of the forestructures of understanding in relation to what he called the "hermeneutical circle."[17] According to this latter concept, understanding any part of our world depends on a prior connection with or preunderstanding of the whole, and any understanding of the whole can proceed only from an understanding of, or projection from, the parts. This circle is not vicious, because we are already in the midst of our life-worlds with certain operative prejudgments. These are either fulfilled or modified as we go on, leading us to learn about and deal with the world. Our anticipatory ideas guide us so that we are not blind; at the same time, by becoming conscious of and criticizing these forestructures we can check the tyranny of the hidden.[18]

Normally, in the process of understanding, the hermeneutical circle expands concentrically.[19] In interpreting phenomena it is crucial to open new meaning by uncovering still-efficacious meanings from the past that bear on the present in ways that have been concealed by naturally shifting intermediate horizons (that is, over time) or by partial and derivative meanings that have come to act as blinders, restricting and monopolizing our focus.

In an influential exercise that is itself an example of a hermeneutic rereading that proceeds by uncovering changed assumptions, Gadamer reinterprets (retrieves) the meaning of *prejudice* in the Western tradition and shows how historiography and other objective methodologies share, without acknowledgment or awareness, a prejudice against prejudice. He analyzes how the original meaning of *prejudice,* of which our current cultural understanding is a diminished derivative, is that of Heidegger's "forejudgment."

We see a transition in this concept with the dawn of the modern era and Descartes's writings. Prior to the Renaissance prejudgments were seen not as false but as the source and bearer of authority and dignity. One could live coherently and learn because one could assume that tradition provides access to truth and positive connection to reality. The modern critique of authority and tradition, however, took acceptance of preexistent authority as opposite to the newly desired certainty that was to be tested by radical doubt and based on no foundation except the self-conscious subject's clear and distinct ideas. The project for radically "objective," self-founding knowledge thus entailed that decisions or judgments previously considered legitimate because grounded in tradition's authority subsequently came to be seen as hasty or loose. Reason itself, as a seemingly ahistorical process, was established as its own and the only authority.[20]

Subsequently, the romantics reversed the Enlightenment's prejudice: the origin, the original, became privileged, and the ancient was taken to have greater import than the present or future progress and perfection. The modern era's idea of primeval stupidity was replaced by the romantics' idea of primeval wisdom, and progress was rejected in favor of the view that civilization is a loss of meaning or regress of mind.

Gadamer powerfully shows how the same structure obtains in both cases while the placement of the elements is inverted in a kind of mirroring: in both cases a prejudgment is made for or against the power of tradition, of authority, based on the *historical* relation to the "original." For the Enlightenment, only what ahistorical reason shows to be possible or impossible, true or false, can be understood in history; for romanticism, reason is replaced by the whole of the past, so that the contemporary can be understood only in the light of its relation to the past, that is, in terms of a universal and radical historicism.[21] The Enlightenment and romanticism are alike in breaking from the older assumption that meaning occurs within a freely taken on, living tradition where prejudgments provide continuity and the stable basis for validity. The Enlightenment and romanticism differ in their prejudice against prejudice by arguing— presuming a change of access to meaning—about whether we have come to see more or less in comparison with the original situation.[22]

Gadamer's point is that we need to remove the prejudice against prejudice, since we are in a tradition whether we like it or not. The very idea—and tradition—of an objective, ahistorical knowledge "belongs, in fact, to historical reality itself," as does historicism.[23] Gadamer argues that historical and cultural research can accept the prejudgment that tradition can operate with authority and dignity, avoiding the need to begin each investigation with radical doubt and stingy criteria for the evidence of the senses. In other words, we can distinguish legitimate prejudices from those to be overcome and thus approach built forms and cultures within their traditions and contexts. By acknowledging that we belong within history and always stand within some tradition, where authority is recognized and accepted in an act of reason and freedom, not in blind obedience, we do not frustrate understanding but open ourselves to it.[24]

In the case of understanding the built worlds of other peoples and times, we begin by recognizing that their traditions are sources of meaning, even if those people are unaware of it. The meaning of what people make always goes beyond the makers' deliberate intentions, because the makers are acting in a manner that involves not only what they consciously intend to accomplish but also their taken-for-granted or unconscious cultural attitudes and responses to the world. People act and think in the context of their historically based insertion in the world, which constitutes the historical reality of our lives more than do our individual judgments.[25] One is little concerned with the individuality of the author and focuses instead on the many dimensions of a life-world, such as shared assumptions, which necessarily remain unspoken and unthought for those within the specific horizon of time and place. What we do always has more meanings and implications than we intend or even can understand ourselves.

Because of this open polysemy, it is possible—even inevitable—that others can find meanings in texts and works not apparent at the time they were made and can learn about shared concerns across time and space in ways that never could have been anticipated. Hence, hermeneutics rejects the attempt made by E. D. Hirsch, Jr., and others to resuscitate the privilege of the author's intention.[26]

Because our horizon is finite and changing and because surplus meanings are available, temporal distance is not a separation or gulf to be bridged; instead, it supplies the ground of the process in which the present is rooted.[27] Through time we are connected to the concerns and problems that earlier people had. We do not seek the exact knowledge of what others thought (to know it as well as or better than they themselves did), somehow occupying their context only and not our own. Nor do we stay merely within our cultural context, alien from any other. Rather, at times we manage to go beyond the limited historical context of either situation, arriving at a widened or comprehensive context where we share something of importance with the other culture and may come to a new understanding of the cultural forms in question, an understanding that may help us to deal with our contemporary problems.

Precisely because past conditions differ from those of the present, the consideration of those differences can be fruitful. In tracing out the connections we ought not try to combine just anything with preexistent, stable meanings; instead, differences between the past and present allow what was hitherto autonomous to be newly combined in us, so that we can have a new experience of meaning. Clearly, despite misconceptions that suggest this, hermeneutics is not at all nostalgic, for it seeks not past meanings but novel combinations of past and present that can occur only in us.[28]

When our spheres of concern and those of others intersect, new possibilities of meaning are opened. Through the investigation of the others' accomplishments, our situation might allow us to see things in previous worlds that had been missed; through the broadening of our context, we might see more possibilities for our situation than we would have if we had remained in our original, narrower context. The other tradition and ours become simultaneous. Here we arrive at a *fusion of horizons.* Gadamer holds, then, that understanding is always the fusion of horizons, where the past and present contexts come together to make something new of living value.[29]

Heidegger's major contribution to hermeneutics may lie in his insistence that the environment and things, texts and language, are not primarily epistemological phenomena, as the modern age would have it, but ontological. He argues that we preconceptually are immersed in a life-world, that texts, political acts, and built things are the catalysts for the disclosure of the world. Hence, words and things are not signs of a prior, independently existing reality, as the tradition holds, nor are they the endlessly self-circulating deferrals that deconstruction admits. Rather, interpretation operates at the scene of the disclosure of our worlds, so that hermeneutics is concerned with the recovery of meaning in the sense of being the occasion wherein new meaning is experienced.

A case in point is Gadamer's previously mentioned retrieval of previous notions of prejudice, which enables us to see that meaning is not something that we can produce as we will but rather a dimension of and event within the shared historical realms in which we and our interpretations belong. In fact, we belong in our life-world primarily through the processes of understanding.[30] The human situation consists largely in acts of interpretation that allow us to belong within our life-world and to experience changes.

Since works of all sorts—bridges and texts and declarations of independence—are what bring a world to stand, setting it into work, the hermeneutic question is about the modes in which particular acts, events, and things are bound up in the appearance and concealment of historical worlds. For hermeneutics, *world* denotes not the collection of all entities but how we are disposed to and within them, that is, the historically disclosed mode of meaning and life. As we have seen, things do not have a fixed, preassigned meaning; rather, the variable meanings of all the dimensions (things, texts, historical worlds, humans) occur or are gathered simultaneously in the event of the life-world.

Given the continual, complex, and plural generation of open-ended meanings, hermeneutics aims to be open to the way the subject matter questions us (our assumptions and views) and attempts to come to understand what the world requires of us as an adequate, appropriate response for participation in its historical unfolding. Thus, as the immediate goal of interpretation—and here we see the hallmark of hermeneutic procedure—we seek not to "create" meaning but to "remove hindrances so the event of understanding can take place in its fullness and the work can speak to us with truth and power," that is, ontologically.[31]

Deconstruction

Whether because it is too new to be a developed tradition or because it is so eccentric, deconstruction is dominated by one figure—Jacques Derrida—in a manner unparalleled by either the traditional approach or hermeneutics. Hence, explaining the approach substantially means explaining Derrida's theory and practice, although others increasingly are using his strategies. Derrida initially accepts the accomplishments and moves of hermeneutics against the tradition, but he goes on to push them to their extremes, eventually employing them against hermeneutics itself to open up a distinctively more radical attitude. His essay "Restitutions of the Truth in Pointing [*pointure*]" is something of a tour de force against the assumptions and practices of both hermeneutics and traditional scholarship. Here Derrida simultaneously criticizes Heidegger's ontological exposition of a life-world that supposedly is set into work by the peasant's shoes painted by Van Gogh and art historian Meyer Schapiro's scholarly identification and attribution in regard to the shoes. (Schapiro, "who claims to hold the truth of the shoes [of the picture]," argued in correspondence with Heidegger that the shoes actually are those of the city dweller, Van Gogh himself.)[32]

Agreeing with hermeneutics, Derrida opposes the traditional assumption about the independent, objective status of things, events, and meanings. He wants to destroy the traditional belief that we can transcend experience or our texts to contact something "autonomously there." For example, he says that reading "cannot legitimately transgress the text toward something other than it, toward a referent (a reality that is metaphysical, historical, psychobiographical, etc.) or toward a signified outside the text whose content could take place, could have taken place outside language."[33] Deconstruction holds that there is no objectively transcendent reality, no essences of things, no clear, stable, or decidable identities. Because these things do not exist, it is impossible to have any direct intuition of or access to them.

This rejects the metaphorical understanding of our language and of signs as transparent, as directly connecting us to their referents. Hence, Derrida consistently criticizes clarity and transparency, not to promote obscurity, but to resist the idea that signs function as "picture windows," to use Northrop Frye's phrase.[34]

In rejecting both the naïve and traditional versions of realism, Derrida opposes the "unperceived or unconfessed metaphysics" and epistemology that provide the bases of scientific objectivism (that is, the metaphysical view of objective reality and its correlate view that truth is a correspondence between language and that reality).[35] This position appears in many variations, as the denial of presence, identity, linear history, causality, and truth. In one of his analyses of Mallarmé's texts, Derrida counters the traditional approach's development of themes in the attempt to work out the truthful correspondence of a work to the world and creator that it supposedly mirrors: "What we will thus be concerned with here is the very possibility of thematic criticism, seen as an example of modern criticism, at work wherever one tries to determine a meaning through a text, to pronounce a decision upon it, to decide that this or that is a meaning and that it is meaningful, to say that this meaning is posed, posable, or transposable as such: [what] have systematically been recognized [as such] by modern criticism . . . cannot in fact be mastered as themes or meanings."[36] That is, one may pursue or trace out themes within the literary sign systems, but that does not allow one to transcend the text or signs:

> If there is a textual system, a theme does not exist (. . . "no—a present does not exist . . ."). Or if it *does* exist, it will always have been unreadable. This kind of nonexistence of the theme in the text, this way in which meaning is nonpresent, or nonidentical, with the text . . . [recognizes that variable meaning] already prevents a theme from being a theme, that is, a nuclear unit of meaning, posed there before the eye, present outside of its signifier and referring only to itself, in the last analysis.[37]

By thus rejecting the foundations that uncritical realism provides for traditional scholarship, Derrida also opens the way for opposing hermeneutics, which in many ways agrees with much of deconstruction's critique of objectively real, determinate meaning. Derrida, however, goes on to contend not only that there is no objective meaning available outside language but also that the cultural-historical processes constituting a world as analyzed by hermeneutics do not involve ontological events, or shared worlds, but instead amount to nothing other than systems of signs and absences. He asserts that there are no foundations or starting points for participation in a historically meaningful, common realm, nor are there any conclusions or essential goals to be reached or any objective, stable accomplishments, such as retrieving original meaning to disclose new dimensions of a world present today. "You find yourself being indefinitely referred to bottomless, endless connections and to the indefinitely articulated regress of the beginning, which is forbidden along with all archaeology, eschatology, or hermeneutic teleology. All in the same blow. '*The new text without end or beginning*'..."[38]

Derrida also disagrees with hermeneutics concerning the plurality of meaning. Hermeneutics characterizes the pluralism of meanings as a freedom from the tradition's preoccupation with stable, univocal, or literal meanings; deconstruction, however, finds not a plurality of meaning but only misunderstanding and failure of meaning. Instead of discovering polysemy, then, deconstruction finds no meaning, since there always is the delay and deferral of meaning, while signs (inescapably) indefinitely refer to one another.

Although deconstruction and hermeneutics agree that there is an endless happening or openness of meaning (versus the tradition's view that meanings become fixed or objectified), deconstruction denies the hermeneutic view that meaning occurs within horizons, that is, within the expectation of transcendent meaning or anticipation of coherence that constitute and proceed from a shared tradition. Derrida "repudiates the assumption of inevitable orientation towards meaning" since it depends on endless depth of text or inexhaustible meaning; rather, "any specification of meaning can only function as a self-defeating attempt to stabilize and restrain what he terms the 'dissemination' of the text. Meaning is not retrieved from apparent unmeaning, but rather consists in the repression of unmeaning."[39]

Derrida no longer accepts the idea of horizons of understanding, that is, of a loosely bounded historical context in which meanings are achieved and of the possibility of discovering new meanings in a live tradition of fused horizons that we share over time. Derrida says, "If dissemination, seminal *différance,* cannot be summarized into an exact conceptual tenor, it is because the force and the form of its disruption explode the semantic horizon."[40] He thus opposes the hermeneutic "stress on multiple, or even infinite meanings, [because it] still attempts to evade this rupture."[41]

Since deconstruction holds that there is no shared, common understanding (or world) at the end of our interpretations, Derrida opposes metaphors of and belief in "depth," and "recovery," which mistakenly perpetuate the ideas of meaning and the purported historical, ontological events of revelation and concealment. Instead, he promotes and speaks in terms of "surface," "play," and "undecidability." Derrida contends that the stabilization of meaning that we achieve results only from the arbitrary preferences and impositions carried out by regimes of power and ideology. There is no fulfillment; at best, texts challenge the assumed worlds of interpreters.

There is no intuitive self-presence of subject and object prior to the representation that occurs in language, in the signs or traces that constitute the world. Rather, he contends, all that signifies or is signified necessarily involves or results from processes of mediation. The mediation that we experience occurs via signs, whether in their primary form as words and language or in other forms such as drawing. The signs move in a flow of repetition, connecting the past to the future and simultaneously undermining the immediacy of the self-presencing. Consequently, the present (what is present and the empty space between past and future) is, strictly speaking, an illusion.

Although we have the sign, and it is all that we can actually have, the sign gives us not any transcendent and present other but only absence. The other is given in the sign, but it is given as absent. How else would the sign represent? The signified is precisely what is not there, is not anywhere; there is no presence, not even elsewhere. Although the sign moves toward the "transcendent signified," it is impossible to complete the trajectory.

We find only traces or *différance* prior to positing and pretending presence or identity. As Derrida uses the term, *différance* (with an a̲) indicates the inescapable difference between what we have (i.e., the sign) and what remains absent (i.e., the signified), between the identity or indiscernability that we wish we had and the nonidentity or discernability that actually obtains. In a further complexity Derrida notes that the crucial "verb 'to differ' [*différer*] seems to differ from itself. On the one hand, [as noted], it indicates difference as distinction, inequality, or discernability; on the other, it expresses the interposition of delay, the interval of *spacing* and *temporalizing* that puts off until 'later' what is presently denied, the possible that is presently impossible. Sometimes the *different* and sometimes the *deferred* correspond (in French) to the verb 'to differ.'"[42] The same ambiguity is maintained and cultivated in the English transliteration *difference* when used in deconstructive theory and practice.

Neither self-identity nor presence ever occurs, since in the mediation of signs and representation, signs always temporally delay and spatially remove the referent, giving us a gap and the absence of what is referred to rather than any coincidence. This is unavoidable. It is impossible to close off the differing and deferral.[43] The possibility of truth, simultaneously, amounts to the possibility of disappearance. For Derrida, what we take to be true is only a fictive construction, a fabrication produced by *différance,* the primary generator.[44]

Because absence permeates time and space and because signs mediate all our experience, repetition and memory are vital for us. They attempt to fill in the gap, the absence. Although they cannot succeed, they are our only resources, our only strategies in the midst of absence. We have no choice but to go on. As a result of *différance,* and in the midst of the reign of *différance,* we and language can and do go on by suppressing absence and positing presence and identity. Usually unconsciously, we conceal the primal *différance,* implementing our preferences within that difference by deploying our choices, privileging one dimension of *différance* over the other.

That there is no direct apprehension of any "transcendental signified" motivates Derrida to interpret drawings in terms of images of blindness and the mediating paraphernalia we try to use to see. Another example of the unavoidable *différance* of identity, presence, and truth is found in Derrida's analysis of "we" and "us" in *Margins of Philosophy.*[45] Heidegger, in developing hermeneutics' ontological-epistemological assumption of necessary prior understanding, claims, in effect, that "*we always already* conduct our activities in an understanding of Being."[46] But who are "we"? Derrida's strategy is to undo "humanism and a certain truth in Being" by exposing both of them as produced and indissolubly tainted by our arbitrary privileging of certain global, cultural, and ethnic differences. "It automatically follows, then, that this *we*—however simple, discreet, and erased it might be—inscribes the so-called formal structure of the question of Being within the horizon of metaphysics, and more widely within the Indo-European linguistic milieu, to the possibility of which the origin of metaphysics is essentially linked."[47]

Thus, for Derrida, nothing is clear and distinct or characterized by untroubled identity. Sexual identity, truth, being, presence, the "author" and "work of art," and so on are all fictions constructed within language and *différance*. This is to say that insofar as we set out to determine what woman, truth, I, or whatever really is, we run up against *différance* that does not go away, except as we ignore it. *Strictly speaking, what anything might be is undecidable.* We belie, or pretend to belie, undecidability by supplementing what is given, substituting or adding other views, interpretations, desires, and beliefs. This again shows the bed of language within which we labor: language, with its thousands of texts, provides the supplement that we use to push the undecidable into being decided.

It is implied in the schema of alternatives, whether of binaries or the dialectic of three, that the meaning and identity of each term must be decidable, but they never are.[48] Decoding terms in the whirl of texts shows that the "crucial experiment" does not exist and cannot be construed. To provide a prime example, Derrida arrays a history of woman and truth by way of a history of an error.[49] Rereading Nietzsche, Derrida elaborates how woman is understood both positively and negatively, as identical with and opposite to masculine truth and untruth: "The question of woman suspends the decidable opposition of true and non-true and inaugurates the epochal realm of quotation marks which is to be enforced for every concept belonging to the system of philosophical decidability. The hermeneutic project which postulates a true sense of the text is disqualified under this regime. Reading is freed from the horizon of the meaning of truth of being, liberated from the values of the product's production or the present's presence."[50]

Nothing is decidable because everything remains within the play of signification, where signs refer only to other signs within the infinite system of signs. The traditional Western desire for and goal of univocal meaning is necessarily barren, except that in its efforts at insemination it engenders delirious illusions of literal meaning and phantoms of identity. Against the hermeneutic hope for polysemy, the profligate and desperate production of seed amounts only to endless loss and spillage, never fruitful result: "Dissemination is the state of perpetually unfulfilled meaning that exists in the absence of all the signifieds."[51] In *dissemination/dissemenation* seed falls only to barren ground.

Derrida projects an image that combines the history of the

Western project and the effective capture of that history within the infinite web of infra-referential signs by playing with Plato's seminal analogy of the cave, inverting that image, turning it inside out and on itself like a Möbius strip.

> Imagine Plato's cave not simply overthrown by some philosophical movement but transformed in its entirety into a circumscribed area contained within another—an absolutely other—structure, an incommensurably, unpredictably more complicated machine. Imagine that mirrors would not be in the world, simply, included in the totality of all *onta* and their images, but that things "present," on the contrary, would be in them. Imagine that mirrors (shadows, reflections, phantasms, etc.) would no longer be *comprehended* within the structure of the ontology and myth of the cave—which also situates the screen and mirror—but would rather envelop it in its entirety, producing here or there a particular, extremely determinate effect.[52]

These examples provide a key to much of Derrida's vision and deconstructive strategies: the way that we experience the necessary mediation through signs that undermine self-presencing and harbor absence coincides, for deconstruction, with the necessary interpretive strategy—what is given is always given so as to undo itself. The search for certitude in our cultural undertakings undoes the very projects that set it into motion. Silence and death are prime examples of the intentionality that projects us toward completion but winds up uniting us with absence, undoing our project.[53]

In interpretation, speaking of *"différance,"* the "undecidable," and "dissemination" are ways of saying and working out our situation within the system of endlessly inter-/infra-referring sign systems that have no outside. In such a universe (our universe, according to Derrida), "The meaning of meaning . . . is infinite implication, the indefinite referral of signifier to signified. . . . [Its] force is a certain pure and infinite equivocity which gives signified meaning no respite, no rest, but engages it in its own *economy* so that it always signifies again and differs."[54] "There is nothing before the text; there is no pretext that is not already a text":[55]

> There has never been anything but writing; there have never been anything but supplements, substitutive significations which could only come forth in a chain of differential references, the "real" supervening, and being added only while taking on meaning from a trace and from an invocation of the supplement, etc. And thus to infinity, for we have read, *in the text,* that the absolute present [and] Nature, . . . have already escaped, have never existed; that what opens meaning and language is writing as the disappearance of natural presence.[56]

Where no sign carries meaning in itself, meaning occurs only within systems of selection and arrangement, which amounts to privileging some possibilities and suppressing or marginalizing others. With nothing outside the realm of signs, nothing outside our texts, the only possible procedure is to expose the illusions of the tradition and proceed positively by joining in the playful movement within the unfolding system of fictions. Our situation and deconstruction coincide as our only nondeluded prospect.

Since all meaning is delayed and deferred and since everything is *undecidable,* there is no point in seeking the chimera of "objective" meaning, as traditional approaches do, or shared polyvalent meanings, as does hermeneutics. The theoretical interpretation of our situation provides the only sane strategy: we need to learn to play by learning to disrupt and to meld intertextuality. That is why Derrida takes up the play of texts and images of representation and why, for example, in treating the drawings of the blindfolded in relation to Plato's allegory of the cave, he can speak of himself as "arbitrarily interrupting this infinitely echoing discourse."[57]

Joining the play must be done by interrupting because the dominant Western tradition has deluded itself about its accomplishment and reality: we have to play along, but we can play along, holding our own in the textuality, only insofar as we "deconstruct the 'illusion' or 'error' of the present."[58] The project is to expose what is claimed to have presence, identity, and truth, to disrupt the exclusions that are in force, that is, the privileges that are sincerely and naïvely claimed to be legitimate. "The break with this structure of belonging can be announced only through a *certain organization,* a certain *strategic* arrangement which, within the field of metaphysical opposition, uses the strengths of the field to turn its own stratagems against it, producing a force of dislocation that spirals itself throughout the entire system, fissuring it in every direction and thoroughly *delimiting* it."[59]

Of course, this means that deconstruction is unavoidably skeptical and ironic. Meanings are given and actions undertaken, for otherwise, life would not go on. But the meanings and actions are not founded in any transcendent reality or knowledge; the situation is ironic since irony simultaneously posits and undercuts what it posits. We can only invent style and meanings as needed, "producing the fictions we need" even while believing in nothing.[60]

Not surprisingly, Derrida targets the Western discourse about representation and art as the site for his disruptions and productions, since here the metaphysical and epistemological assumptions swirl about most powerfully. He is particularly interested in slipping into the central texts already underway, Plato's and Kant's theories of representation and art—Plato's because it is the original and master text in a long series and Kant's because he reposits for the modern era the distinction between art and reality that needs to be undermined. Derrida criticizes the distinction and seeks to overthrow the supposed dichotomy of either/or.

Derrida's deconstructions of the graphic and visual arts illustrate this procedure. Throughout his writings Derrida critiques the ordered line because the line is the sign of causality, history, identity, presence, linear logic (discourse as *dis-cursus,* that is, as running on to a conclusion), and drawing. Graphics, whether generated by writing or drawing, are a major factor in our technics of cultural illusion and suppression.[61] "If there is no extratext, it is because the graphic—graphicity in general—has always already begun, is always implanted in 'prior' writing."[62] Although the signs try to trace out lines of successful external connection, Derrida wants to show how they always centripetally fall back into themselves.

In short, because we operate within the texts and intertracings that are always already in effect, the linear assumptions of time, causality, identity, logic, and so on, on which the tradition depends, have to be given up. It is just this arbitrary nonfoundation that the tradition fails to recognize and that must be brought to the fore and erased by deconstruction. Philosophy and the other disciplines depend on this model of linearity and are yet oblivious to it. "The enigmatic model of the line is thus the very thing that philosophy could not see when it had its eyes open on the interior of its own history. This night begins to lighten a little at the moment when linearity—which is not loss of absence but the repression of pluridimensional symbolic thought—relaxes its oppression because it begins to sterilize the technical and scientific economy that it has long favored."[63]

Given the project of exposing the groundless foundations of metaphysical theories and the signs/graphics of representation, art, reality, presence, history, identity, and causality, it is not surprising that Derrida combines these elements in his *Memoirs of the Blind.* The very models of lines that we cannot see even as we look at ourselves and our history need exposure. Because we need to become skeptical, blindness—the loss of direct perception and intuition—has its heuristic uses. It is our blindness that we must come to see, which we can do by deconstructing drawing and the tradition of representation, including the entire cultural history of environmental construction and interpretation, with its successive failures of unmediated vision. Thus blinded, in the midst of the hidden *différance* and forgotten tracings on which we depend, we and our environments are exposed—or need to be exposed.

The Book's Organization

As noted, the book does not follow the usual procedure of moving through arguments or positions to arrive at a certain conclusion. It more simply presents examples of three very different ways of interpreting three kinds of environment. Because there is no implied order of best and worst, how should I array them in a linear text that unavoidably presents something first and something last? Given the historical development and principled differences among the theories, their arrangement could be other than it is. The basic explanation for the present arrangement is that it displays the dynamic among these live, contested alternatives.

I put the tradition first simply because it is the "base" from which we start and with which the reader is likely to be the most familiar, and because it is the "foundation" that the other two theories reject or would replace. Even Derrida notes that you have to have the tradition before you can move on.[64]

Deconstruction comes next because it accepts the position—the positing—of the tradition and aims to undo it. The deconstructive interpretation proceeds by first laying out traditional readings of the environment (by Sigfried Giedion and Richard A. Etlin) only to show how that tradition undercuts itself. Moving from traditional scholarship to deconstruction, we have a smooth flow and, perhaps, accumulating understanding.

Hermeneutics appears in the third chapter because it tries to steer a middle course between tradition and deconstruction. (That is not to say that it does or, even if it did, that a middle course somehow is preferable to one of the two extremes. This issue is part of what is contested among the theories.) Still, since the hermeneutic interpretation of American nature does position itself between the traditional and the deconstructive readings, it makes sense for it to follow them. In addition, in the postscript, I explicitly contrast the hermeneutic example with traditional and deconstructive readings. The reader is free to decide for him- or herself.

Again, the order could be otherwise. Hermeneutics partly claims to be more radical than deconstruction and yet also more conservative—more of a middle way and yet deeper. Deconstruction rejects depth and is not impressed by any position claiming to be close to tradition or to retrieve it. As for the tradition, it holds that both deconstruction and hermeneutics are extreme and mistaken in rejecting objective meaning. Each approach would draw its own map of the respective placements. The task here is much simpler: to present the complex interpretive terrain so that its topography remains intelligible and accessible and the readers' choices of paths stay open.

Although it might be instructive to read the same environment in three different ways to make a systematic comparison, it also would be boring—and unnecessary, since the alternative approaches themselves often "unread" their rivals as part of their own procedure, and because, along the way, I treat comparable, alternative works so that readers can follow the differences if they wish.

It bears saying that all three approaches could and do apply to all three sorts of environment examined in this book. It is no doubt possible to argue that a particular method fits a particular sort of environment best, but it seems to me that the tradition, deconstruction, and hermeneutics are powerful enough to apply throughout. So, why not some diversity, for fuller coverage of the environments that matter to us and for variety's sake?

The chapters treat three different sorts, or scales, of environments: houses, public spaces and monumental buildings, and the American landscape. Although there is no hard-and-fast reason to divide the environment into these three realms, doing so gives us an overall coverage of the cultural and environmental subject matter. The result, I hope, is to stimulate us to think about the entire, complex built world that we inhabit.

Both ordinarily and in the research literature, we discern, normally without devious or subtle connections in mind, the immediate spheres of our private lives, the local, public spaces and monumental buildings that provide a civic realm, and the larger "natural" landscape that becomes culturally shaped into the sphere within which our settlements are inserted and operate. There is thus a kind of obvious progression of scales if we want to consider a range of environments, and it makes sense to start by considering the smallest, which perhaps will be the easiest to manage.

To develop the environmental subject matter more fully and to provide the opportunity to raise additional questions, I consider an increasingly broadened scope for the human realm of these three environments, moving from the individual over a lifespan, through groups that historically have defined themselves in terms of specific ideologies, to groups attempting to develop an enduring national "identity" over the course of hundreds of years by means of complex political and religious or secular belief systems. I make this final variation more to complete the possibilities than to suggest any correlation between groups and environments, much less between groups and interpretive approaches. Again, it seems obvious and benign enough to begin with the relation of individuals to their houses and proceed first to particular groups in public spaces at specific places and times and then to heterogeneous groups of people in a landscape attempting to become a nation across several centuries.

I present these variations not to work out systematically the three approaches across three scales of built environment, with three human scopes, but to enrich our examples and experience of thinking without intruding on the basic problem, strategies, and outcomes that characterize each interpretation. That is, we can have the variations without obscuring what it is that makes the interpretations alternatives.

Chapter 1, then, moves back and forth between biographical and cultural information on the one hand and houses on the other. This is meant to provide a baseline, a clear example of the traditional approach. Ludwig Wittgenstein's and Carl Jung's life stories and their cultural contexts are woven together with their theoretical views and the houses they designed and built and with what appear to be the basic alternatives for interpretation and action in the world.[65]

Chapter 2 takes up deconstruction. The traditional account and the supposed "reality" of Egyptian, French neoclassic, and postmodern pyramids, where the built environment mirrors cultural belief systems and a kind of identity and presence is achieved, is seen as an illusion and failed fabrication. By showing how both the pyramids and their cultural projects—postures—unavoidably undo themselves, deconstruction seeks to unmask and defer untenable postures and adopt in their place strategies that explicitly posit arbitrary order.

Chapter 3 applies hermeneutics as practiced by Mircea Eliade, Hans-Georg Gadamer, and Martin Heidegger. The hermeneutics of American nature nonnostalgically aims to uncover disclosures that occurred, among other places, in nineteenth-century landscape painting and landscape design. The interpretation shows how the originally twinned religious meanings of American nature as a paradise given and as a negative wilderness to be overcome are now concealed and forgotten, although in their secularized transformations they still operate in our understanding of parks and wilderness and in our attitudes to the use and development of land. With the recovery of the tradition of non-self-willful attitudes to the landscape, new possible meanings may emerge in the fusion of Native

American and ecological thinking.

I hope that all the analyses help to make the approaches' strategies and applications more accessible and that they help the readers to work out their own positions. I also hope that the environments discussed shine forth as newly problematic, as no longer taken for granted or forgotten. Insofar as that happens, the book might stimulate a new interest in and zest for almost all the world and challenge us to a less smug and judgmental attitude, a new tolerance for and understanding of differences.

At the same time, it is not likely that these contending approaches and praxes can in a particular case be equally appropriate as responsible ways to conduct research and shape our increasingly shared environments or that the interpretations they yield can be simultaneously "true." Choices will be made, and the choices matter. The book aims, then, not only beyond theory to the practice of environmental interpretation but finally toward interpretation as the heart of ways of living, where our choices among powerful alternatives are bound up with our future and the world's.

Chapter **1**

Traditional Approaches

Wittgenstein's and Jung's Lives, Work, and Houses

1

Traditional Approaches
Wittgenstein's and Jung's Lives, Work, and Houses

Facing Uncertain Meanings and Traditions

Suppose we want to become at home with ourselves, our families, and the world around us. In what ways is it possible to develop our individual and social identities adequately and properly? In what ways is it possible to understand and build meaningfully and valuably? The problem is especially pressing in our contemporary situation because of our uncertainty about both tradition and the future. Our experience since at least the turn of the century and through two world wars shows that science and technology, our history, philosophy, mythology, and domestic and urban environments, which continue to shape us, are as dangerous as they are powerful, as full of illusion as of promise. How then to live within our culture in a manner that is neither bitter nor nostalgic, neither deluded nor unnecessarily impoverished?

One way to approach an answer is to look at exemplary individuals' personal lives, professional work and theory, and the houses they built. For example, Ludwig Wittgenstein and Carl Gustav Jung both pursued the problem of who we are and how we should build, that is, of how to become at home; worked and built as a "therapeutic" and analytic means to articulate the meaning of language, symbols, thought, and existence; and arrived at a hard-won, yet profound simplicity and at houses that are monuments to who they were.

Their difference, however, as seen in their personal lives, work, and houses, is fundamental. Moreover, although each worked out a viable mode of becoming at home, the two alternatives may exhaust—or at least represent—our only contemporary possibilities. Wittgenstein strove to clear away the misleading dimensions of the past and to clarify affairs by reducing problems to their basics and then objectively ordering those fundamental elements. The congruent style uncompromisingly faces cultural and personal illusions and our confusions about meaning and value. It austerely accepts that there is no final solution, no rest.

In contrast, Jung worked to enrich and manifest complexity by recovering layers of meaning from the past and allowing contradictions to become manifest. He encouraged deep and obscure symbolism to come to consciousness and undergo transformation so that its lessons can be integrated with the "responsible" empirical consciousness. Such transformation helps to guide the gradual process of individuation, which patiently moves toward a final unity and peaceful wholeness.

Wittgenstein's Restlessness

Ludwig Wittgenstein (1889–1951) remained restless all his life, moving among strikingly different—if not opposite—modes of existence, pursuing each one for a time with almost total preoccupation and in an "excited state." He studied physics in Berlin, aeronautical engineering in Manchester, and, seized by fundamental issues, formally took up philosophy at Cambridge. Next came a year of solitude in Norway. The following year, 1914, found him a volunteer in the Austrian army, deliberately seeking dangerous assignments and working toward positions nearer and nearer the front.

After the war Wittgenstein rejected his family's wealth, which he gave to his brother and sisters, and took teacher training in Vienna, followed by a series of assignments as a country schoolteacher (in Trattenbach, a tiny mountain village, and then Ottertal and Puchberg at Schneeberg). He next worked as a gardener's assistant at Hütteldorf and the seminary at Klosterneuberg, until he broke off to spend two years building a mansion for his sister Gretl. The house completed, he went back alone to Norway and thence to Cambridge again, where he became a professor at Trinity College after an extraordinary examination. Since he had no formal terminal degree in philosophy, he was questioned by his eminent supporters, Bertrand Russell and G. E. Moore, about his seminal work, *Tractatus Logico-Philosophicus,* which served as his thesis even though it had been published eight years earlier. Thereafter Wittgenstein pursued philosophy in a brilliant but anguished manner.

In this varied and intense life's course, two constant features and patterns emerge: a very strong and even driven personality and an apparently irresistible attraction to the contrary demands and claims of the technological and the philosophical (especially to philosophical problems of mathematics, logic, language, and psychology). Because of the severity with which these projects gripped him, the intense concentration he needed for each task at hand, and the disparate natures and requirements of the different undertakings, Wittgenstein suffered a good deal. He was so tormented, for example, that he was unable to stop thinking and fall asleep unless he first lost himself in a movie.[1]

In many ways Wittgenstein's life became almost entirely his work, since the work so preoccupied him. Although working—thinking, writing, teaching, and building—did not relieve his agitation in the end, at least it provided a realm for the exercise of his bottled energy and resulted in "products" satisfying enough to provide a feeling of conclusion, freeing him to move on to another task.

Wittgenstein's philosophy reflects his restlessness and the severe, rigorous demands of mathematics, logic, and technique. Indeed, his overall career is marked most obviously by his two major works. The *Tractatus* influenced significant movements in the philosophy of science and language (logical atomism and Vienna Circle logical positivism). Although these traditions continued on their own, both were later rejected by Wittgenstein himself in his second book, *Philosophical Investigations.*

The *Tractatus,* in its austere and simple style, aims at logical clarification and shows that philosophical thought is an *activity,* not a body of knowledge or set of theories.[2] That is, philosophy clarifies misunderstandings that result, for example, from mistaken views of language and from conflations of the sensical and nonsensical. Wittgenstein believed that it is easy to be misled by false analogies and images. Consequently, as the undoing of specific knots and problems, philosophy is a kind of analysis or therapy on language and thought.[3]

At the same time, however, the *Tractatus* takes on a strange kind of objectivity, a textual autonomy enhanced by the serial order of its often enigmatic remarks. The whole work thus conveys the single image of a conceptual world.[4] For instance, in dealing with the complex relationships among linguistic statements, propositions, and assertions, the *Tractatus* notes the "logical space" in which a proposition occurs:

> 3.4 A proposition determines a place in logical space. The existence of this logical place is guaranteed by the mere existence of the constituents—by the existence of the proposition with a sense.
>
> 3.41 The propositional sign with logical coordinates— that is the logical place.
>
> 3.411 In geometry and logic alike a place is a possibility: something can exist in it.
>
> 3.42 A proposition can determine only one place in logical space: nevertheless the whole of logical space must already be given by it. . . .
> (The logical scaffolding surrounding a picture determines logical space. The force of a proposition reaches through the whole of logical space.)[5]

Here we see the twinned aspects: (a) a set of remarks that have sense only as used in the context or process of clarifying and unmasking a specific problem and (b) a detached set of sayings that strangely and powerfully stand on their own and speak for themselves.[6]

Wittgenstein's later work is very different, albeit with some similarities. *Philosophical Investigations* has very little by way of logical sequences of ideas or lines of thought strung out for readers to see and follow. Rather, the book is a composite of remarks, sayings, observations, and notices.[7] It has been compared to a travel diary, where the order of what is said reflects the itinerary of the journey across the landscape, where partial views and multiple sketches describe (without explaining) the encountered features of that landscape. If the *Tractatus,* then, is a map providing a single image of a conceptual world, *Philosophical Investigations* is an album of varied landscapes and sketches.[8]

Consistent with the early view of the *Tractatus,* the later Wittgenstein continued to hold that philosophy helps to keep us from being misled and is an analytic therapy useful for undoing problems. Differing with his earlier view, however, in *Philosophical Investigations* Wittgenstein emphasizes remaining bound to (or returning to) ordinary language as a means to avoid theory building and "creative" thinking. This move enables us to stay down to earth, where what is encountered is left alone and clearly shown so that we can understand what appears as problematic. As a means to escape from traps, blind alleys, and confusions in thought and language, Wittgenstein's remarks are to be used and then left behind.

And Wittgenstein the architect? Between bouts of (with) philosophy, Wittgenstein spent two intense years (1926–1928) designing and overseeing the building of his sister's residence, an activity she intended to be relief and therapy for his tormented state after World War I.[9] Wittgenstein, although not formally trained as an architect, had an engineering background, as we have seen, and in the process of securing the necessary permits and directing the building of the house—later the Bulgarian Embassy Monument and now again accessible—he signed official documents as "architect." For the two years that Wittgenstein gave himself entirely to the project (characteristic of his way of life), he and his sister agreed that the only matters *not* of concern were time and money. (Margarete Stonborough-Wittgenstein belonged to one of the wealthiest families in Vienna.) The preoccupation with and control of even the minutest details, then, resulted from Wittgenstein's personality and disposition and from his sister's resources and support.

The site, at Kundmanngasse 19, was a mixture of the unusual and common. The neighborhood and surrounding houses were simple and anything but cosmopolitan. The property itself, however, was special: a 33,000-square-foot former horticultural nursery set high

Figure 1.1. Stonborough-Wittgenstein House from the south.

above the street and therefore apart from its immediate surroundings. The property contained an old house, as well as a small garden and beautiful old trees.[10]

Initially, in the spring of 1926, the architect Paul Englemann was commissioned to sketch the Wittgenstein residence. In the fall of 1926 Margarete Wittgenstein asked her brother to participate. He soon took over the project completely, largely as the result of his strong personality and "uncompromising demands."

The site's blend of the ordinary and unusual is reiterated in the contrast of the mansion's exterior and interior. In conformity with the neighborhood of unprepossessing houses, the cubic exterior is not especially striking (see fig. 1.1). The modern exterior design, however, is somewhat strange in that Wittgenstein thought very little of most modern architecture—"though he admired, for example, van der Null and Adolf Loos, whose work Wittgenstein House somewhat resembles."[11] Wittgenstein felt alien from the "main current of European and American civilization . . . manifest in the industry, architecture, and music of our time" and would not accept "what nowadays passes for architecture,"[12] since, for example, "in modern architecture they don't know in what style to design a building."[13] In any case, clearly the exterior is "of its time and place, of its culture," perhaps an overriding consideration since it allowed an appearance consistent with what we know of Wittgenstein: he took pains in his style of life to appear unpretentious and ordinary on the outside, whether as a gardener or boarder, although he burned with energy and excitement within. Congruently, the interior of the house is anything but modest. As might be expected, in the words of one architectural critic, the house "is unique in the history of 20th. century architecture"; within, "everything is rethought."[14]

Several features are worth special attention: the windows, doors, floors, exposed technological fixtures (lighting, elevators, radiators), and the overall spatial effect. The windows and glass doors witness Wittgenstein's preoccupation with proportion and technical control, making almost impossible demands on materials and craftsmanship (see fig. 1.2). The windows and doors exhibit the significance of proportion, a feature most important to Wittgenstein, as can be gauged by his later reaction (in 1930) to his rented quarters at Cambridge, "where he had chosen rooms at the top of the staircase, . . . [and] altered the proportions of the windows by using strips of black paper."[15] Wittgenstein commented, "See what a difference it makes to the appearance of the room when the windows have the right proportion. You think philosophy is difficult enough, but I can tell you it is nothing to the difficulty of being a good architect."[16] Not surprisingly, when his sister's house was finished, the one feature Wittgenstein was not satisfied with and wanted to change was a set of three windows on the second floor.

The glass doors and windows were incredibly difficult to construct because of the size of the glass panes and because the iron dividers between the panes have no horizontal support. Indeed, so uncompromising was Wittgenstein that he rejected the first completed door after waiting several months for it. Eventually the negotiating engineer became hysterical, staying on only because of the commission and his own professional standards. Wittgenstein later conceded that of the eight firms he negotiated with, only the one would have been able to meet "what I had to demand" in "precision and objectivity."[17] In an even more extreme demand, Wittgenstein insisted on raising one room's ceiling 1¼ inches after it was completed and the house was ready for the final cleaning!

Figure 1.2. Stonborough-Wittgenstein House: windows, glass doors, columns.

The doors reflect the same preoccupation with proportion and detail, for example, in the handles and locks designed individually by Wittgenstein. The material is important, since metal permits total control and, with a clear lacquer covering, has a highly "finished" surface, both features contributing to the severe effect (see fig. 1.3). The cut and polished stone of the floors functions the same way.

Especially striking are the exposed elevator and light bulbs. The technological is shown just as the logical and ordinary modes of language are in Wittgenstein's philosophy: frankly acknowledged and let be. The lighting is uncompromisingly austere and "honest" about its appearance.

The two small (originally) black cast-iron radiators tell the whole

Figure 1.3. Stonborough-Wittgenstein House: door handles designed by Wittgenstein.

story in miniature. Wittgenstein's efforts on them, from design to initial delivery, spanned an entire year. Severe technical difficulties were met in material selection and preparation: each of the radiators is made of two parts that stand at precise right angles and meet in the corner, with a tiny space between them. The result was made more difficult because the radiators rest on legs that had to fit exactly to produce the precise spacing. Consequently, the elements had to be cast outside Austria and then ground to meet the specifications to the millimeter. The final symmetry of the two smooth black objects across from each other in a small room results in a precise and proportioned form. The flawless simplicity and rigor are so characteristic of sculptural beauty that the radiators were entirely

10

congruent with (and, on occasion, even served as bases for) the Chinese artworks that were a prominent aspect of Gretl Wittgenstein's furnishings when she lived there.[18]

The overall spatial arrangement and effect are intimately bound with the axial-symmetrical lighting. Placement and connection of homogeneous, symmetrical spaces, each with the same orientation and value, result in a static order. Against this homogeneity and its attendant monotony, Wittgenstein counterbalances "significant irregularity" by means of the subtle differences and variations among doors, windows, and fixtures and the more noticeable recessed column heads. The architectural details function as "phenomena akin to language in music or architecture."[19] Finally, then, the dominating impression of static symmetry has a painstakingly worked-out counterpoint, although that too is brought within the formal unity of composition.

Although evidently a fine setting for Gretl and congruent with her personality, which was also bent to a combination of the ordinary and unusual, the house appeared to their sister Hermine, through its perfection and monumentality, as a "dwelling place for the gods," not people, this "house turned logic."[20]

This assessment of the house is quite compatible with Wittgenstein's view on architecture, the point of which, he felt, is to be monumental. "Architecture immortalizes and glorifies something. Hence there can be no architecture where there is nothing to glorify."[21] He also held that architecture is gesture: "Remember the impression one gets from good architecture, that it expresses a thought. It makes one want to respond with a gesture."[22] Evidently architecture, like language, is not something other than, in addition to, or foreign from thought but is itself the vehicle of thought.[23] But what was the gesture of Wittgenstein House? What was the thought? The answer has several dimensions.

First, the house was a gesture showing the culture in which it arose. Wittgenstein was as severe a cultural critic as he was a philosopher of language. "I once said, perhaps rightly: The earlier culture will become a heap of rubble and finally a heap of ashes, but spirits will hover over the ashes."[24] Just as the linguistically misleading, confusing, and nonsensical are to be cleared away, even more so are the sham, phony, and bankrupt dimensions of society.

Although Wittgenstein does not speak of the Bauhaus in the available documents, obviously his is the same response as that of Loos and the Bauhaus: the tangled web of falsifying, posturing, and dangerous Western "symbolism" and propaganda had to be stripped away and the no longer valid dimensions of tradition rigorously eliminated. As Wittgenstein noted, "Today the difference between a good and a poor architect is that the poor architect succumbs to every temptation and the good one resists it."[25] The resulting austerity must be accepted and left to stand, rather than covered over immediately by fictive, falsely comforting symbols and ideologies newly invented to take the place of the old. Stark honesty is better.

We see, then, that Wittgenstein House corresponds to Ludwig Wittgenstein's philosophical thinking, and both cohere so that at least at one point, at one place along the way, Wittgenstein articulates something approaching a unified self. Wittgenstein's lifelong intensity and uncompromising personality are satisfied (partially, at least) by the clarity and rigor that we see carried out in the reduction, simplicity, and ordering not only of sentences but also of windows and doors. The same precision and (attempted) control appears in the logical sequence of thought and in the metallic doors and windows. Wittgenstein sought polished ideas and architectural elements alike. The austerity of the simple, spare logic and house leaves a final impression of detachment. With its objectified proportion and the relation of materials, the house has a stark, autonomous beauty—not unlike Zen koans, which appear enigmatic and, simultaneously, carefully crafted to be exactly as they are: spare and precise.

Here, again, philosophical thought and building coincide.[26] The general effect of the mansion—the composition of homogeneous, connected spaces, apparently value-neutral since no direction is favored or orientation emphasized—is one of objectivity and self-containment, an appearance parallel to the single vision of the *Tractatus* and Wittgenstein's autonomous vision at the time. As the *Tractatus* lays out the logical space of propositions, Wittgenstein House lays out the physical space of Wittgenstein's "statement."

Finally, however, the house and *Tractatus* are merely milestones. Although we occasionally produce what become detached objects and ideas (in the form of a house or philosophical volume), philosophy and architecture really are processes, akin to therapy that seeks intelligibility. This congruence between Wittgenstein's life of philosophy and architecture, both as activity, is witnessed by remarks in which he speaks of one in terms of the other, remarks especially significant since his professional specialty was uncovering and clarifying misleading analogies.[27] In 1930 he noted that civilization constructs, that progress "is occupied with building an ever more complicated structure" and hence is concerned with clarity only as an instrument. "For me on the contrary, clarity, perspicuity are valuable in themselves. I am not interested in constructing a building, so much as having a perspicuous view of the foundations of possible buildings."[28]

Later, in speaking of the importance of unexpected views for the final use of philosophy, Wittgenstein remarks, employing the same language that he uses in the *Philosophical Investigations,* that "a man will be *imprisoned* in a room with a door that's unlocked and opens inwards; as long as it does not occur to him to *pull* rather than push it."[29] As philosophy and architecture illuminate each other, they also belong to the activity and mode of one's life. Acknowledging the conjunction, Wittgenstein notes, "Working in philosophy . . . like work in architecture in many respects . . . is really more a working on oneself. On one's own interpretation. On one's way of seeing things. (And what one expects of them.)"[30]

Philosophical insights and remarks are to be used to solve a problem and then left behind as we move on. The Wittgenstein villa also is to be left behind. Like the distressed culture it monumentalizes, the house disclosed that it lacked what it needed for a full, satisfied life—a goal all too illusory not only to Wittgenstein but to contemporary society: "The delight I take in my thoughts is delight in my own strange life. Is this joy of living?"[31] In 1940 Wittgenstein assessed matters thus: "the house I built for Gretl is the product of a sensitive ear and *good* manners, an expression of great *understanding* (of a culture, etc.). But, *primordial* life, wild life striving to erupt into the open—that is lacking. And so you could say it isn't *healthy* (cf. a 'hothouse' plant)."[32]

Wittgenstein was restless precisely because, in addition to being sensitive, mannered, and understanding, he ultimately was bound to a wild dimension of existence. So Wittgenstein moved on, back

Figure 1.4. Ludwig Wittgenstein, Cambridge, 1946. Photograph by Dorothy Moore; reprinted by kind permission of Mr. Timothy Moore.

to philosophy. The house, like the *Tractatus,* was only one partial and temporary view of things. Of course, we should not forget that the house was Gretl's, not Ludwig's, and that it suited her.

When Wittgenstein moved on, what home did this restless personality occupy? Nothing more than the barest of rented rooms, where the interior, filled with his burning and tormented self, may have had a quality, like Van Gogh's rooms, that the verbal description alone fails to capture (see fig. 1.4):

> Wittgenstein's rooms in Whewell's Court were austerely furnished. There was no easy chair or reading lamp. There were no ornaments, paintings, or photographs. The walls were bare. In his living-room were two canvas chairs and a plain wooden chair, and in his bedroom a canvas cot. An old-fashioned iron heating stove was in the centre of the living-room. There were some flowers in a window box, and one or two flower pots in the room. There was a metal safe in which he kept his manuscripts, and a card table on which he did his writing. The rooms were always scrupulously clean.[33]

Jung's Quest for Wholeness

Carl Gustav Jung (1875–1961) was the first child in the family of a "poor country parson" in northern Switzerland; from that rural world he went to *gymnasium* (high school) in Basel, a situation that yielded embarrassing encounters with wealthy classmates. Although he decided that he wanted to study science, in those times he would not have been able to support himself as, say, a zoologist, so he "compromised" by pursuing a medical career. He completed medical school and his examination at the University of Basel and in 1900 began psychiatric work at Burghölzli, the mental hospital of Canton Zurich and the psychiatric clinic of Zurich. He finished his M.D. dissertation in 1902 and remained at the clinic. Later he studied in Paris, married (eventually having five children), began his correspondence with Freud (in 1906), and in 1909 made his first trip to the United States—with Freud.

Freud provided Jung with professional support and referred to him as the "crown prince," although they eventually broke over professional differences. In 1909 Jung moved to a house in Küsnacht that he helped to design and focused on private practice. Jung's remarkable series of publications worked out his epochal ideas in analytic psychiatry: the personal and collective unconscious, individuation, animus/anima, persona (all begun by 1916), the "self" and psychological types (by 1921), and archetypes and symbolism (1917–1918). In 1922 he purchased property in the village of Bollingen, and in 1923 the first tower was built there. Jung worked and lived alternately in Küsnacht and Bollingen until he died.

Jung's own lifelong journey to selfhood was intertwined with his psychological concerns and houses, both imaginary and concrete. As a child he played at an old stone wall in the family garden, carrying out little rituals (such as tending fires) in the "caves" formed by the interstices of the blocks. He also secretly carved a little wooden figure of a man that he kept with a painted stone on a roof beam in the forbidden attic of his house and attended with ceremonial acts (such as adding little scrolls of writings). Jung tells us that the secret comforted him, so that he "felt safe, and the tormenting sense of being at odds" with himself disappeared.[34]

As Jung grew up house images played an important part in his development. While a schoolboy he became aware of two conflicting aspects of his personality, which he labeled number 1 and number 2: number 1 (the main personality) was scientifically empirical and concretely oriented; number 2 (a "shadow" personality) was intuitive and inclined to the historic and fantastic. Images of houses wound through both dimensions. At times he gave way to the number 2 dimension in systematic fantasy, reveries about an island like a hill of rock, where "on the rock stood a well-fortified castle with a tall keep, watchtower. This was [his] house."[35] In contrast, when his number 1 aspect later found such "fantasy silly and ridiculous," he turned to building and studying castles and fortified encampments.[36]

In Jung's lifelong struggle to unite these two contradictory dimensions—with the eventual understanding that the split was neither actual nor peculiar to himself but intrinsic to a dynamic "played out in every individual"[37]—dreams involving houses and built environments played a persistent and decisive role. For example, consider one of many dreams closely related to his work that Jung had in the second half of his life. In 1928, in association with painting a mandala with a golden castle in the center, Jung dreamed of being with companions in Liverpool, where they found a square on a drizzly, foggy night. The square had a round pool in the center, and in the middle of that was a small island. Although the surroundings were obscured by rain, smog, and smoke, the center was brightly lit, and in it a single magnolia tree blossomed.

Apparently none of his companions saw the focal light and tree. Jung comments, "This dream brought with it a sense of finality. I saw that here the goal had been revealed. The center is the goal, and everything is directed toward that center. Through this dream I understood that the self is the principle and archetype of orientation and meaning. Therein lies its healing function."[38] After this realization Jung gave up drawing and painting mandalas, since the dreamed center satisfied him by giving "a total picture of my situation."[39]

Similarly, in 1912 Jung had confronted the unconscious (that is, the idea of the unconscious) in dreams. He dreamed of his previously mentioned childhood projects of building little houses and castles of bottles and stones with mud mortar, which had taken place when he was ten or eleven, some twenty-six years earlier. The dream's unusual emotional power made Jung realize that the activity was still important to him and that it posed the question of how to unfold a creative life. Only after great resistance and with hard-won resignation did the established, esteemed psychiatrist resume what appeared to him to be the only approach to develop this important dimension in himself, again taking up his childhood games of building. The reluctance is understandable, "for it was a painfully humiliating experience to realize that there was nothing to be done except play childish games." Jung explains:

> I began accumulating suitable stones, gathering them partly from the lake shore and partly from the water. And I started building: cottages, a castle, a whole village. The church was still missing, so I made a square building with a hexagonal drum on top of it, and a dome. . . .

> I went on with my building game after the noon meal
> every day, whenever the weather permitted. As soon as
> I was through eating, I began playing, and continued to
> do so until the patients [of his private practice] arrived;
> and if I finished with my work early enough in the
> evening, I went back to building. In the course of this
> activity my thoughts clarified, and I was able to grasp
> the fantasies whose presence in myself I dimly felt.
>
> Naturally, I thought about the significance of what I
> was doing, and asked myself, "Now, really, what are you
> about? You are building a small town, and doing it as if
> it were a rite!" I had no answer to my question, only the
> inner certainty that I was on the way to discovering my
> own myth. For the building game was only a beginning.
> It released a stream of fantasies which I later carefully
> wrote down.
>
> This sort of thing has been consistent with me, and at
> any time in my later life when I came up against a blank
> wall, I painted a picture or hewed stone. Each such expe-
> rience proved to be a *rite d'entrée* for the ideas and
> works that followed hard upon it.[40]

Perhaps the best example of a dream of a house as rite of passage
and entrance is found in one of Jung's dreams that Freud attempted to interpret during their visit to the United States and that, after
Freud failed, led Jung to the idea of the "collective unconscious." In
the dream Jung found himself on the second floor of a house with
a salon well furnished and decorated in the rococo style; going
downstairs he found everything older, from the fifteenth and sixteenth centuries, with medieval furnishing and red brick floors.

Exploring the house he came on a heavy door, which opened to
a stone stairway leading to the cellar. Descending further Jung
found himself in a beautifully vaulted room that was very old, with
walls dating from Roman times and a floor of stone slabs. In one of
the slabs was a ring; by pulling on it he lifted the stone slab beneath,
again uncovering a "stairway of narrow steps leading down into the
depths." Going down Jung finally entered a low cave cut into the
rock, where in the dust he found bones, including two human

skulls, and scattered pottery "like the ruins of a primitive culture."

Although Freud's inability to show Jung the meaning of the dream, despite a pretense to do so, helped to confirm the differences that led to their break, the significant point here is Jung's own interpretation: "It was plain to me that the house represented a kind of image of the psyche—that is to say, of my then state of consciousness, with hitherto unconscious additions. Consciousness was represented by the salon. . . . The ground floor stood for the first level of the unconscious, . . . the cave [for] a world which can scarcely be reached or illuminated by consciousness . . . that borders on the life of the animal soul."

The rooms signified past times and surpassed stages of consciousness, which Jung later came to understand as archetypes: "My dream thus constituted a kind of structural diagram of the human psyche; it postulated something of an altogether *impersonal* nature underlying that psyche . . . The dream became for me a guiding image . . . [and] was my first inkling of a collective *a priori* beneath the present psyche."[41]

Houses, then, were intimately connected with Jung's self-realization and professional work, but not only in dream and image, in childish construction. The two houses where Jung spent his adult life clearly manifest his self-individuation and represent his ideas and the two dimensions of his personality.

The family residence and site of Jung's private practice at Küsnacht, just outside Zurich, was designed by Jung in collaboration with his cousin, architect Ernst Fiechter. Set on the water's edge, it was designed in the style of the old farmhouses of Canton Zurich.[42] Jung had a motto carved above the door: VOCATUS ATQUE NON VOCATUS DEUS ADERIT ("Summoned or not, the god will be there"). The motto is the answer that the Delphic oracle gave to the Lacedaemonians who were planning a war with Athens. In Jung's explanation, it says, "yes, the god will be on the spot, but in what form and to what purpose? I put the inscription there to remind my patients and myself that 'the fear of the Lord is the beginning of wisdom' (Psalms III:10)."[43] In addition, while in England, Jung had carved wooden reproductions of the little wooden manikin he had made and hidden in the attic as a child; he now had a larger version made in stone and placed in the garden at Küsnacht.

It is the house at Bollingen, however, that most magnificently bodies forth the path of Jung's individuation, specifically, "the path by which a person becomes a psychological 'individual,' that is, a separate, indivisible unity or whole."[44] In Jung's words:

> Gradually, through my scientific work, I was able to put my fantasies and the contents of the unconscious on a solid footing. Words and paper, however, did not seem real enough to me; something more was needed. I had to achieve a kind of representation in stone of my inner-most thoughts and of the knowledge I had acquired. Or, to put it another way, I had to make a confession of faith in stone. That was the beginning of the "Tower," the house which I built for myself at Bollingen.[45]

The site is near water on the upper lake of Zurich in the area of St. Meinrad; the property was once church land, belonging to the old monastery of St. Gall. (The connection with religion and the dead is not accidental, although it is beyond my scope here.)[46] Jung bought the land in 1922 and built the house in stages from 1923 to 1935, after which he continued to add stone carvings, sculptures, and paintings to the interior walls. The house at Bollingen primarily was Jung's private retreat, although he did entertain visitors on occasion, signaling their welcome by lowering the flag that indicated "solitude only."

Initially Jung had planned a primitive, hutlike dwelling, a round structure surrounding a central hearth (connected with the wholeness of the family), which even domestic animals share in some societies. Such a building would have been too primitive, however, and thus destabilizing to Jung's own direction and consciousness.[47] Instead, the first structure, although round, was "a regular two-story house." Substantially and simply built of stone, it was a "maternal hearth" and a "suitable dwelling tower" (at the left of fig. 1.5).[48]

Although Jung felt rejuvenated during his stays at Bollingen, he also came to feel that something was lacking. So, four years later, in 1927, he added the large central structure, with its towerlike annex (see fig. 1.5). Then, after four more years, the building still seemed too primitive and incomplete, "so in 1931 the tower-like annex was extended." Jung said, "I wanted a room in this tower where I could exist for myself alone."[49] The idea was to have a space for personal withdrawal, as Indian houses do for meditation or yoga. He continued, "In my retiring room I am by myself. I keep the key with me all the time; no one else is allowed in there except with my permission. In the course of the years I have done paintings on the walls, and so have expressed all those things which have carried me out of time into seclusion, out of the present into timelessness. Thus the second tower became for me a place of spiritual concentration."[50]

A desire then arose in Jung for a larger, open space where he could be connected to the sky and nature. Consequently, again after four years, in 1935 he fenced in a portion of the land by building a courtyard and loggia by the lake. Thus, a fourth element was added "that was separated from the unitary threeness of the house," making the whole quaternity in four-year segments over the course of twelve years (see fig. 1.6).

Finally, in the congruence between Jung's self-realization and Bollingen Tower, after his wife's death in 1955, when he was eighty, Jung made the last addition (see fig. 1.7). He tells us:

> To put it into the language of the Bollingen house, I suddenly realized that the small central section which crouched so low, so hidden, was myself! I could no longer hide myself behind the "maternal" and "spiritual" towers. So, in that same year, I added an upper story to this section, which represents myself, or my ego-personality. Earlier, I would not have been able to do this; I would have regarded it as presumptuous self-emphasis. Now it signified an extension of consciousness achieved in old age. With that the building was complete.[51]

Life at Bollingen was of the simplest form, establishing connection with the timeless rhythms of nature and the heritage of the symbolic past (see fig. 1.8). Jung explains:

> There is nothing in the Tower that has not grown into its own form over the decades, nothing with which I am not linked. Here everything has its history. . . . I have done

Figure 1.5. Bollingen: dwelling tower, central structure, and annex, 1927. The estate of Carl Jung.

Figure 1.6. Bollingen: courtyard, two towers, and loggia, 1935. The estate of Carl Jung.

Figure 1.7. Bollingen: completed, with upper story in center, 1955. The estate of Carl Jung.

> without electricity, and tend the fireplace and stove
> myself. Evenings, I light the old lamps. There is no run-
> ning water, and I pump water from the well. I chop the
> wood and cook the food. These simple acts make man
> simple; and how difficult it is to be simple! [52]

As a help to work out and enter the meaning of his life and the tower, Jung, who was reputed by local craftsmen "to know his stone," worked on a series of stone carvings. In 1950, while building the enclosing wall for the garden, Jung encountered a square stone—intended for a cornerstone, but entirely the wrong shape and size as shipped from quarry—into which he carved a Latin alchemical verse ("Here stands the mean, uncomely stone, / 'Tis very cheap in price! / The more it is despised by fool, / The more loved by the wise"). On the front face, in the stone itself, he discerned a small circle, or "eye," which he chiseled in and to which he added a small homunculus and a Greek inscription concerning the child's cosmic play ("Time is a child—playing like a child—playing a board game—the kingdom of the child. This is Telesphoros, who roams through the dark regions of this cosmos and glows like a star out of the depths. He points the way to the gates of the sun and to the land of dreams").

On the third side, which faced the lake, he added more Latin alchemical inscription, in which the "stone speaks for itself" ("I am an orphan, alone; nevertheless I am found everywhere. I am one, but opposed to myself"). Jung says that he wanted to chisel the phrase "Le cri de Merlin" into the back face (fourth side) of the stone, but did not, because the stone's message reminded him of Merlin's life in the forest: according to the legend, Merlin's cries are still heard, but no one today can understand them, and thus his story remains unfinished. The stone was placed outside the tower and meant to explain it: "It is a manifestation of the occupant, but one which remains incomprehensible to others."[53]

> From the beginning I felt the Tower as in some way a
> place of maturation—a maternal womb or a maternal
> figure in which I could become what I was, what I am
> and will be. It gave me a feeling as if I were being reborn

Figure 1.8. Jung tending fire at Bollingen, 1949. The estate of Carl Jung.

in stone. It is thus a concretization of the individuation process, a memorial *aere perennius*. . . . Unconsciously built at the time, only afterward did I see how all the parts fitted together and that a meaningful form had resulted: a symbol of psychic wholeness. . . . At Bollingen I am in the midst of my true life, I am most deeply myself.[54]

> In the Tower at Bollingen it is as if one lived in many
> centuries simultaneously. The place will outlive me, and
> in its location and style it points backward to things of
> long ago. There is very little about it to suggest the pre-
> sent. If a man of the sixteenth century were to move into
> the house, only the kerosene lamp and the matches
> would be new to him; otherwise he would know his way
> about without difficulty. . . . It is as if a silent, greater
> family, stretching down the centuries, were peopling the
> house. There I live in my second personality and see life
> in the round, as something forever coming into being
> and passing on.[55]

The tower at Bollingen helped the unconscious and collective to
emerge, then, and complemented the house at Küsnacht, which was
the primary scene of consciousness and action, exemplified in the
responsible treatment of patients. Together the two houses strength-
ened each dimension of Jung's personality and made an opening for
the integration that they now memorialize.

Alternatives for Contemporary Existence

For all their differences, Jung and Wittgenstein paralleled each other
surprisingly in the way they built houses in an effort to realize their
identities. Their houses, then, are monuments to their alternative
"archetypal" responses to the possibilities of contemporary existence. Both felt dual and conflicting forces or attractions in their
personalities, especially between the empirical and concrete (Jung
with science and Wittgenstein with technology) and the historical
and theoretical (Jung with his humane mythology and symbolism
and Wittgenstein with philosophical remarks and fragmentary sayings). Both spent enormous energy and concentration on building
and tending an intense, often painful and distressing personal life by
working out new styles of analysis and therapy (in psychiatry and
philosophy, respectively) and by building as a way to reconcile the
self's dimensions and needs. For Jung and Wittgenstein, building
was the focus of great care, exactitude, and involvement and an aid
to a significant relationship with their familial and cultural contexts.

Their buildings are monuments to their modes of living in the face of uncertain meaning and tradition. Wittgenstein House and Bollingen Tower each achieve a powerful simplicity. Both are direct and sparing in material and form. Wittgenstein's glass, metal, polished-stone flooring, exposed elevator and light bulbs, and homogeneous and symmetrical space all resulted from a demanding control that enabled them to "remain standing as themselves" in the single vision of the moment. Jung's rough stone and plaster walls carved and painted with symbols, as well as the heterogeneous and "directionally charged" elements and the absence of any technology, even electricity and running water, made an opening for the gradual manifestation of and integration with the timeless. Wittgenstein's house and life stand austere, as a clarified objectification; Jung's, as a rich, almost tangled primal mystery.

Finally, for both Jung and Wittgenstein, the completion of the crucial house marked an accomplishment, a conclusion. For Jung, the achievement was the matured emergence of the second dimension of his personality. Bollingen and the second, or shadow, personality complemented Küsnacht and the first personality. Building the house made possible the completion of his life's course and self-wholeness and also the arrival in a timeless realm where he could dwell in participation with ancestors, family, nature, and place. For Wittgenstein, building his sister's house was the completion of a phase that opened to the next swing, to philosophy and life in a starkly bare room. His action was a stage in the continuing movement of a restless, unsettled life. In large measure, these two patterns of living by building for serene dwelling or restless movement remain the best alternatives we have today.

With Wittgenstein we are challenged to courageously face culture's misleading and falsifying elements and to advance clarity relentlessly. In short, we starkly stay within the ordinary and the hard surfaces of the way things are, not in misty heights or obscure depths. Wittgenstein shows us that we can be at home while remaining unsettled and that a house is a monument to the activity of building, which appears as a vital part of our sequences of movement. With Jung we see the task of integrating conscious understanding and action with the unclear, confusing, and disconcerting contents of the unconscious into a totality of the personality, that is, into the self. This process requires a willingness to risk the depths of the past, the unconscious, the nonpersonal, and the nonsensical to find the possibility of archetypal power and self-unity. Jung shows a way to come home to one's own self, our life's goal, which also is the path to the self's settling into and dwelling within the "timeless" past and an enduring symbolic cosmos. Here houses not only concretize the individuation process and its rites of entrance but also, and finally, manifest and memorialize that completion.

Chapter **2**

Deconstruction

Pyramids as Posture and Strategy

2

Deconstruction

Pyramids as Posture and Strategy

> The silence of prehistoric arcana and buried civiliza-
> tions, the entombment of lost intentions and guarded
> secrets, and the illegibility of the lapidary inscription dis-
> close the transcendental sense of death as what unites
> these things to the absolute privilege of intentionality in
> the very instance of its failure.

—Derrida, *Husserl's Origin of Geometry*[1]

Deconstructing Pyramids

The pyramids. We scarcely notice the way we refer to or think of them. That the word can stand alone, without qualification, witnesses the power of the name and the built forms over memory and landscape. The name has no need of further specification or modifier. The reference is presupposed and secure: the Egyptian pyramids. There are other pyramids too, of course, those in Central America and, later in the course of Western history, the pyramids of French neoclassicism and the postmodern era. These other and later pyramids, however, refer to the same grounding principles that the Egyptian pyramids witness: the triumph of life over death, the dominance of eternal presence over the fleeting, which passes over into

absence, a sustained personal and cultural identity underlying fractured differences, the realm of the gods and the immortal soul
enduring beyond natural decay and human impotence. In short, the
pyramids refer to the principles of intelligibility and reality that
found Western culture and lay down its goals.

So it seems. Consider, however, a double displacement. First, the
built form we know as the pyramid has for centuries displaced its
antecedents and concealed the fictive basis of its strategies to
achieve, or stand for, a "timeless" presence and identity. Second, the
discourse accepted as the orthodox interpretation has grounded the
cultural meaning of the pyramids in the privileged dimensions or
concepts of presence, identity, and life precisely by suppressing the
correlates (absence, difference, and death, respectively). This posture of domination over one member of each pair attempts to erase
the primal "difference," or gap, at the heart of the relations within
which we are situated. The illusion of mastery cannot succeed. Both
the pyramids as built forms and the orthodox discourse surrounding them (which relates the original foundational acts and sustaining principles) undercut themselves and disclose fissures that reveal
what is hidden: the fictive web spun as the strategy and posture of
cultural forms of desire.[2]

Egyptian Pyramids

We hold onto the meaning of the pyramids, against time and forgetfulness, through discourse (see fig. 2.1). According to the traditional and orthodox interpretation, for example as codified by
Sigfried Giedion, the Egyptian pyramids were the means for entering into eternity, the timeless realm of the gods and life after earthly death.[3] The pyramids, of course, were part of larger complexes,
where all the ceremonies and built elements (temples and tombs,
courts and walkways, storehouses, paintings and reliefs, and utensils
and stores, as well as the pyramids themselves) were there for the
sake of the eternal continuity of the pharaoh's existence. As part of
an intricate understanding of a seven-dimensional soul, the pyramid
complexes derived from the nature and needs of the principal soul,
the *ka*.

The Egyptians believed that the sun god, Ra, was the sole source

Figure 2.1. The means to eternal presence, identity, and life.
Pyramids at Giza.

of the *ka,* the vital force of life. This cosmic, divine force was passed from the sun god to his son, the pharaoh. The sun god was present in the pharaoh, manifest in the human realm; thus, the kingdom was grounded, through the pharaoh, in the cosmos and with the gods. The key to a unified cosmos and to life for the kingdom was the pharaoh's sustained participation in the power emanating from the god. The presence of the vital force ensuring continuity of life was partially accomplished and witnessed by the built environment, specifically, stone architecture.

The final, or entire, function of the pyramids was to hold the pharaoh in an eternal present. Accordingly, they were made of polished stone, the most indestructible substance for the dwelling of the *ka* (see fig. 2.2). They had closed chambers filled with symbolic and actual means of sustenance that radiated a brilliant and powerful beauty.[4] The form itself replicated the rays emanating from the sun (the vital force descending to earth). Originally the capstones were covered with gold, which would gather, convey, and celebrate the sun's color and light, thus connecting god, pharaoh, and land, a connection focused by the pyramid's alignment to the cosmic quadrants. As one of the pyramid texts prayed: "*Atum,* so put thine arms . . . about this pyramid, as the arm(s) of a *ka,* that the *ka* of [the pharaoh] may be in it enduring for ever and ever."[5]

Thus, a series of binary terms founded the pyramids. The dominant terms (eternal presence, life, sky, sun, etc.) were meant not only to dominate but also to deny and overcome their mates (absence, death, earth, and moon).

As texts, however, the pyramids contain elements displacing themselves, that is, undercutting the meaning that was so long supposed.[6] Eternal presence is not perdurable. The still unity is fractured into multiplicity and succession. Pyramids are found in many sites; pyramids are grouped together. To note only a few such multiplicities: Zoser's step pyramid had six stages; there are the three pyramids of Sneferu, followed by Cheop's and Mycerinus's and the three pyramids of Giza, including Cephren's. Their multiplicity and shared form belie the timeless simultaneity of pharaoh and gods. The total built environment resulted from successive pharaohs, who replaced one another not only throughout the course of serial earthly deaths but also in their possession of the sun god's vital force and,

Figure 2.2. The most enduring material protects the pharaoh's *ka.* Stone pyramid at Giza.

thus, earthly power. The pharaoh's life and conjunction with power, then, were not eternally the same.

Furthermore, the pyramids did not merely contain the absent body of the dead pharaoh but also hid it in layers of symbols and symbolic objects. Indeed, mummifying aims to preserve the body for eternity. That is, mummifying denies or conceals the body and person as dead, as mortal, even though only a dead body is mummified. Although the pharaoh's living soul was to have remained after his bodily death, it appears obvious that he truly was dead: his soul did not continue to dwell in the pyramid. Thus, neither the eternal present nor everlasting life were able to dominate and ground the pyramids and their meaning. Even pharaohs themselves acknowledged the lacuna. Although their own future everlasting life depended on the reality of these sacred forces and beliefs, on occasion they were so bold as to desecrate their predecessors' pyramids or tombs, showing that the power actually did not hold.[7]

Indeed, the pyramids as carefully formed to deal with the absent (dead) pharaoh were insinuated into a culture that pretended it could build so as to obliterate or hide death and time. The culture of the eternally present and enduring life was nothing more than a particular case of, and strategy in regard to, death and absence, for if the pharaoh were not dead and absent, there would be no need or motive for the pyramids. That is, the pyramids testify to the pharaohs' absence and death and their own status as fictive strategy. Similarly, the gods also appeared as an absence never rendered present; they were a fiction derived from and sustained by the immediate built environment of pyramids.

The enduring Egyptian stone gives no hint or traces of its construction and now absent content. The scaffolding, earthworks, and laborers' dwellings that enabled the pyramids to be built have disappeared to let the generated form persist without obvious source and hence without measure or opposite. The forms remain, useless after capstone and contents have been looted and the pharaoh's remains carried away. (In Central America the failure of pyramid complexes to transcend is more obvious: not only have the workers' huts been reabsorbed into the jungle, but so have the pyramids themselves. Although they were more often sacred astronomical observatories than tombs, the American pyramids' facing conceals only rubble, not generative power.)

French Neoclassic Pyramids

Centuries later, when the power of the Egyptian tradition and culture had waned, another—indeed, a doubled—displacement of pyramids occurred. In eighteenth-century France, as part of a neoclassic vision and architecture of death, pyramids again pretended to sustain presence and identity but did not. The French pyramids simultaneously *had to* but *could not* replace the Egyptian. They had to replace the Egyptian pyramids because the ancient forms were no longer efficacious in grounding presence. The unquestionable absence of any guarantee of a triumphant eternal presence needed to be overcome; the lack needed to be filled. Thus, the French tradition had to replace the Egyptian to allow any possibility of presence and identity in the eighteenth century. At the same time, by the very means it employed to displace the Egyptian, the French tradition necessarily undercut itself: the posture and techniques required to overcome the Egyptian immutability unavoidably depended for any power on that very source. In discrediting the Egyptian tradition, the French tradition also unavoidably nullified itself. Thus, the French pyramids dislocated not only the earlier system they would replace but, simultaneously, their only foundation and hope for success.

The French pyramid appeared as part of the gesture through which presence and identity announce their triumph over absence and fracture, that is, as the denial and conquest of death proclaimed in cemeteries and by monuments. This phenomenon in eighteenth-century France may best be focused in the architecture of Etienne-Louis Boullée and in the parallel cultural discourse on death as narrated, for example, by Richard Etlin.[8] Perhaps better than any of his contemporaries, especially in his drawings published from 1782 to 1799, Boullée completed the appropriation of the Egyptian pyramids and their meaning to develop for the sake of the living a new form for the cities of the dead. In the project of delineating the neoclassic vision of life's victory over death, Boullée and his colleagues drew on and fused the two principal manifestations and underlying traditions of the sublime: the sublimity of nature, whereby we can experience the timeless, and the example of great individuals, which in commemoration allows an intimation of immortality.[9]

According to this concept of the sublime, the immensity and

overwhelming features of nature can induce an experience where the mind is led to behold divinity. At base this was a development of the traditional Christian idea that creation bears witness to its creator, but in the modern age the creator was no longer understood as wholly other than the created, as he had been in the tradition of negative theology. Rather, divinity was held to be present in nature itself. In the French pantheism connected with this concept of the sublime, nature was seen as the appearance of the timeless.[10]

Death, then, was also seen in terms of natural forces, where nature's overwhelming grandeur would move the soul to a relationship with the divine and eternal life.[11] The idea was worked out in the French tradition of the cemetery as Egyptian wasteland, where the fascinating forms of the Egyptian pyramids in the desert were artistically and architecturally represented and reinterpreted during the first half of the eighteenth century. J. B. Fischer von Erlach had drawn them in his *Egyptian Pyramids* and *Theban Pyramids* (1721; see fig. 2.3). His renderings were adapted, in turn, by Jérôme-Charles Bellicard in *Capriccio* (1752) and Hubert Robert with his *Capriccio* (1760). Over the forty-year course of these representations, the pyramids and their surrounding landscapes became increasingly sublime. The initial interpretations, which strongly distorted the pyramids according to the prevailing cultural conventions, gave way to symbolic renderings that deformed the pyramids into shapes evoking mountain majesty, lost in dramatic clouds. The pyramids merged with the naturally sublime and, by identification with nature's timelessness, bespoke—if not quite promised—an endurance beyond death.

In addition, there was a companion phenomenon bearing on the interpretation of death and our alternative strategies to deal with it. By the eighteenth century a French tradition existed wherein cemeteries were understood as memento mori meant to show the public the folly of human vanity.[12] For instance, the cemetery of the Holy Innocents in Paris, by exposing the remains of the many individuals buried there, was intended to provide an occasion for moral reflection and education. Gradually the desire to reorganize human activity physically and to clarify that physical organization provided a motive for new designs of cemeteries.[13] Alongside the concern for health and sanitation, which vigorously emerged as means to dis-

Figure 2.3. J. B. Fischer von Erlach, *Egyptian Pyramids*. From *Entwurff einer historischen architectur,* 1725.

criminate clean from polluted environments, appeared the concurrent desire to separate the morally pure from the corrupt—the basis for cemetery design as it was seen by Blondel and Ledoux.[14]

During this process of new design and meaning for French cemeteries, the concept of monumentality also changed. The earlier tradition of the cemetery as moral lesson against vanity was radically transformed by the humanist tradition of inspiration (which is desire for immortality): the spur to proper conduct became emulation of great achievement. The great deeds of heroes were to be presented to memory by monuments in cemeteries. Indeed, the triumph of lasting fame over death and virtue over vice transcended the cemetery, so that in the 1740s the French began to "envisage the entire city as an appropriate subject for instilling virtue."[15]

Here, too, the pyramids were used as monuments to celebrate the illustrious dead, the heroes whom people should imitate in the course of becoming good citizens. Dufourny, for example, employed a pyramidal central chapel in his *Cenotaph to Henry IV* (project of 1778), thus "associating commemoration with triumph

Figure 2.4. Etienne-Louis Boullée, *Cenotaph in the Egyptian Genre, c. 1785.* HA 55, no. 26. Bibliothèque Nationale, Paris.

in the new cemetery" and showing how sublime architecture is appropriate, as Blondel advocated, "to the sepulchers of great men and in general to all monuments raised to remind our citizens of great deeds, the remarkable exploits, and the value of our princes, heroes, and great generals."[16]

As previously noted, Etienne-Louis Boullée fused these two traditions of the sublime: "On the one hand, Boullée's pyramids are cenotaphs to personal greatness; on the other, they are incarnations of nature itself. The two themes are combined in the great man, who is considered to have reached a status commensurate with the grandeur of divinity."[17]

Boullée drew on these traditions of the cemetery as Egyptian wasteland and the concept of monumentality, but he went on to develop a new interpretation and unity. The heart of his vision was his belief that death is seasonal and that the pyramid is an image of immortality by way of the sublime. Here, contrary to the common view, Boullée did not see the seasons as a cyclical course of birth, growth, decline, and death; rather, he saw a fundamental seasonal opposition: winter was the time of blackness and death, contrary to life.[18]

Boullée understood the pyramid as expressing nature's timelessness. Of course, this was the age of renewed vigor in mathematics,

Figure 2.5. Etienne-Louis Boullée, *Cenotaph to Turenne, c. 1785,* plan. HA 57, no. 23. Bibliothèque Nationale, Paris.

and along with interest in pure mathematical forms came interest in phenomena such as crystalline forms in minerals. "In this work, Boullée may have found a confirmation of his own conviction that perfect geometries were the underlying forms of a universal order. . . . In his *Cenotaph in the Egyptian Genre,* Boullée joined the idea of a return to the bosom of the earth through death with the image of this same *natura naturans* rising out of fertile chaos as a pure crystalline form" (see fig. 2.4).[19]

In a series of major projects Boullée set out how the pyramid and its associated symbols suggest "a return in death to the fertile center of the world in order to join the immensity of the cosmos, which is identified with Divinity."[20] Thus, in his *Cenotaph to Turenne* (c. 1785), a project for a cemetery in Paris, his plan shows how he surrounded the pyramid with an ossuary, which mediated, as a frame, the union of nature and pyramid and set the pyramid as the site of the unity achieved (as an expectation met when the visitor moved to the center) (see fig. 2.5).

In the cenotaph itself Boullée completed the humanistic idea, which had been developing since the end of the seventeenth century, that death can be overcome by human deeds (see fig. 2.6). The fame of heroes, enduring in the monuments and memories of citizens and emulated by the latter, would be a triumph of immortality.[21] Forgoing the established specific imagery, Boullée assimilated "Turenne's valor to a more universal form that could also embody the spirit of nature. Boullée's pyramid, as Jean Starobinski has observed and as the section shows, rises as a *puissance,* a vital force. The pyramid is nature incarnate, timeless, chthonic, primitive, fertile, and crystalline."[22]

Of course, as one of the most venerable monuments known, "the

Figure 2.6. Etienne-Louis Boullée, *Cenotaph to Turenne, c. 1785,* section. HA 57, no. 14. Bibliothèque Nationale, Paris.

pyramid seemed to Boullée to furnish an ideal 'image of immutability.'"[23] The idea of immortality achieved by noble accomplishment was especially forceful when the cosmic order was experienced by the hero. For example, it was believed that in reasoning out the law of gravity, Newton had entered into union with the deepest order and therefore with the mystery of nature, as in other ways had explorers who died in the course of unlocking nature's secrets. Hence, their lavish monuments witnessed the commemoration of their conjunction with nature and immortality.

The emphasis on experience in the encounter with the sublime and the eternal was carried over in architectural forms that allowed the citizen to participate in, or at least catch a hint of, such noble deeds and cosmic grandeur. This was accomplished, for instance, by emphasizing sequential movement over simultaneous visual grasp, as in the "horizontal or temporal progression from the ossuary at the periphery to the pyramid at the center" at the Paris cemetery.[24] Both Boullée's *Cemetery Entrance by Moonlight* (c. 1785; see fig. 2.7), with its dazzling triangle of light at the heart of the pyramid, and its inversion, *Funerary Monument Characterizing the Genre of a Buried*

Figure 2.7. Etienne-Louis Boullée, *Cemetery Entrance by Moonlight, c. 1785.* HA 55, no. 27. Bibliothèque Nationale, Paris.

Architecture (c. 1785; see fig. 2.8), with its darkened triangular center, were designed in a manner that "presented an intimation of immortality at the initial point of entry along the axes to the central pyramid."[25]

Thus, in the French neoclassic architecture of death, and especially as formed by Boullée, we find the triumph of the dominant member of each pair: immortality over death, the timeless over the temporal, nature over individual corruption, great deeds over the shameful (as in the cases of propriety over unseemly behavior or respect over disrespect for social mores), and reason over unreason (as with Newton's penetration of nature's order). Consistently, even rigorously, the age of rationality sought to constitute a culture wherein the mathematical and formal were displayed in the order of social space and in the behavior and manners of the population. Of course, architecture and the built environment not only supported this vision of presence and identity but constructed it.

As noted, the use of pyramids in the eighteenth century architecturally to posit immortality over death depended on the successful appropriation of the Egyptian tradition and the displacement of the original meanings and beliefs. For the Egyptians, the pyramids were identified with the timeless because the pyramids joined the pharaoh and gods, thus overcoming nature's changes and death. Inversely, for the French, nature pushed itself forward as timeless presence, imaged in the form of the pyramid. Whereas nature had been an absence of eternity for the Egyptians, with the French rever-

Figure 2.8. Etienne-Louis Boullée, *Funerary Monument Characterizing the Genre of a Buried Architecture, c. 1785.* HA 55, no. 29. Bibliothèque Nationale, Paris.

sal it took over as the manifestation of the presence of timelessness. The primally divine, as opposite and transcendent to the natural, gave way to divinity identified with (and incorporated into) the natural through the sublime.

Similarly, whereas for the Egyptians the deeds that could confer enduring life were entirely restricted to the gods and pharaoh (thus excluding the rest of the kingdom and giving the pharaoh power over it), for the French humanists the glorious achievements conferring immortality were those of mortal heroes and were open to emulation by the general citizenry. Human reason and action took the place formerly reserved for divinity.

In addition to instituting these grounding displacements, the French asserted their power over the Egyptian in a number of other ways. The mathematics of rationalism replaced nature's order and seemed to explain it fully. Thus, the rational understanding of the pyramid as a pure form or concept destroyed the pyramid understood as a measure or element in Egyptian sacred geometry. Then, too, there is the overwhelming fact that the French appropriation of pyramids resulted from conquest, the by-product of the triumph of reason, which carried away the forms and artifacts (such as obelisks) of Egypt to Western public spaces and museums—the arena for displaying the power of vectorial visual organization and presentation.[26] In both cases the spectacle of the willful, historical use and manipulation of the Egyptian environment and concepts by the

Enlightenment French, which aimed to establish the latter's identity and immortality, makes clear the emptiness of the Egyptian claim to the same triumph.

The French asserted immortality (presence and identity) for ordinary humans by displacing the original gods and pharaoh, an act achieved by overturning the pharaohs' and gods' claims to superiority over natural and human change and impotence. The Egyptians did not achieve eternal life or transcendent immortality but succumbed to time's flow in the forms of changing power and corruption of sham concepts. What had postured as presence became absent.

The suppressed and invisible counterpoint, however, indicates the flaw in the French project and posture: what comes to claim presence when it deposes into absence what had been present was itself initially absent and has no other ground. That is, there is no possible foundation for a claim to immortality, because such a claim's only power would derive from the original forms and concepts that it takes over; but for those original forms and concepts to have a sustained power, they would have to retain their presence and identity—precisely that condition that cannot obtain once they have been rendered absent and different from their purported nature by their French usurpers.

How could the pyramids have facilitated French access to timelessness when their very appearance in French form witnessed the failure of the original Egyptian gods, pharaohs, and people to maintain the primal integrity and desired effect through these forms? To vanquish immortality through architecture the French used the pyramid, which was possible because they displaced the Egyptian system of meaning. At the same time, however, the French could make genuine use of the pyramids only if the pyramids retained their appearance as immutable—although the pyramids could not be immutable unless they, along with the Egyptian gods and pharaohs, retained their original identity and continually manifested their eternal presence in the later French social and natural world, that is, unless they successfully resisted any French appropriation. If such were the case, of course, it would have precluded the French inversion and replacement of Egyptian divinity and sacred power, the inversion necessary for the French appropriation and domination in the first place.

The French attempt for dominion depended on the power of the culture and built environment it necessarily subverted; insofar as the attempt succeeded, it subverted the basis for any sustained power of immutability that inhered in the pyramids, thus becoming impotent itself. There was no triumph, only irresolvable double displacement, as the French inverted the Egyptian system and also thereby their own foundation, in a self-contradiction and self-displacement. It is like attempting to clear a building site with fire and rebuilding, using the still-smoldering lumber, in the same place while the fire rages on, consuming any new building.

This double bind appears to have hidden itself at the time: "The antiquity of the pyramid made it a prime candidate for a monument to convey the immortality of the meritorious dead; . . . as the architect Sobry would express it, in the pyramid, 'immortality . . . is visible and palpable.'"[27] Nonetheless, the very proof cited by the Académie Royale d'Architecture in 1865 as demonstrating the worthiness and venerable age of the pyramids is simultaneously and exactly the evidence that indicates that the pyramids had become mere traces of lost identity, without worth and power: the "absence of hieroglyphs and situation on a sandy plain."[28]

Just as had happened in Egypt, the situation in eighteenth-century France undercut itself. The architectural environments actually *disclosed the fracture* in the supposed appearances of presence and identity, as seen in the displacing inversions at work here. Consider the flaws that historical change makes obvious, betraying the hidden differences usually concealed in the more subtle and powerful struggle of alternative concepts. The pretense that heroic deeds are obvious and endure contained the seeds of its own overturning. Indeed, the humanistic idea of emulating glorious deeds was itself a displacement that inverted the traditional Christian moral: the idea of memento mori was that the vanity of human accomplishments would be clarified in the face of death, whereas the new monumentality raised what had been suppressed (vanity) over its former master concept (humility). So vanity paraded as the key to immortality.

How could the achievements honored for the sake of a dominant class's desire for a more proper order expect to endure in human admiration, much less for eternity? Such views are passing, disputed, and corrupted. Not surprisingly, then, this French vision of immortality through nature and monument itself soon dissolved, being diminished and assimilated by 1800 through the concepts of health and the picturesque (for example, in the cemetery as picturesque landscape garden).

Today that French use of pyramids appears an affected attempt to claim more than is possible: their failure to achieve immortality in union with timeless nature is obvious in the decay of European corpses (an even greater decay than that of the older, mummified pharaoh), and the pyramid form itself, clearly neither derived from nor symbolically attached to a present transcendent realm and divine vital force, was instead the product of self-conscious human artistic and moral efforts.

If the primal gods could not give us immortality, how are we to confer it on ourselves? Surely we cannot do so by conquering earlier cultures and carrying away the meaning of their discourse and built environment. We find a lesson in the silent, and so far ignored, companion of the pyramids at Giza. Since the "Sphinx records the moment of closest identity between god and pharaoh: between the invisible and visible,"[29] the popular tale of Napoleon's artillery destroying the nose of the Sphinx during practice turns out to be a tale of the French spiting their own face.

Postmodern Pyramids

In their recent manifestations in I. M. Pei's project for the Louvre and Arata Isozaki's Museum of Contemporary Art in Los Angeles, pyramids appear as "postmodern quotations."

Pei's glazed pyramid, which is based on the proportions at Giza, is sited in the middle of the Cour de Napoléon at the Louvre (see fig. 2.9). The pyramid, which is surrounded by three smaller pyramids and seven pools, provides ground-level access to the surrounding buildings. The glazed surface, according to the architect's early announcements, "would reflect the skies of Paris by day and be lit like a vast lantern at night."[30]

Traditional methods of interpretation would evaluate the project—controversial, at the least—according to categories such as historical context, formal relation to surrounding architecture, or the creativity of the architect. No doubt "creative," these pyramids

Figure 2.9. Pyramids and pools in the middle of the Cour de Napoléon. I. M. Pei, Grand Louvre Pyramid, 1985–1989.

either display the genius of their designer in the vicinity of one of the major collections of Western art or show the egotism of one who would compare his work with that of the masters. Formally the pyramids appear to shatter the symmetry and unity of their context, resulting in conflict and tension rather than harmony and resolution of rich complexity.[31] Although the fusion of disparate elements can create polysemy, as in metaphor, the mere juxtaposition here appears to be a bad joke.

If we recall the deeper foundations of French art and architecture, however, specifically the mathematical and geometrical basis of great works, whether in David or Cézanne, Pei's pyramid may be seen as more meaningful. The purity of pyramid, cylinder, sphere, and square could be said to be the common basis for "classic" French art, of whatever period and style. Moreover, such pure forms have linked French art to its origin in antiquity (see fig. 2.10). Pei's project may evoke that heritage, too—appropriate for the Louvre, where the focus is on the great classical tradition.

Thus, to make sense, even at a formal level, the pyramids would have to function within the French heritage and development of

Figure 2.10. The pure form of the pyramid and the heritage of "classical" French art? I. M. Pei, Grand Louvre Pyramid, 1985–1989.

form, so that the new project is grounded in the past. The project's intelligibility would be a matter of precisely the displacements and deconstruction previously considered. To transform the issue into the terms I have been using, one would say that the creative dimension is a struggle for presence and immortality. Simultaneously, the historical and formal dimensions contend for presence and identity.

In this interpretation Pei's pyramid functions not merely as a historical quotation of classic tradition and pyramidal form but as the active assertion of presence and identity at the expense of the preceding tradition.

The Mitterrand government, by commissioning a change in the Louvre that appears to fracture its historical repose and aesthetic wholeness, is attempting to assert that what matters is not the conservation of masterpieces from the past, whose physical condition and meanings deteriorate despite our best techniques of preservation. Rather, what matters is the continuous, present creativity that is the French artistic and cultural accomplishment. The Louvre, then, is not an institutional Dorian Gray but witnesses French identity as a creative power actively present.

The bold pyramid, built over the clamor of conservatives who would retain the past and freeze its forms, achieves the active priority of the proper—desired—contradictory dimensions: presence of contemporary creative power and culture versus absence as they recede into the past; the identity of the current regime and art as creative and culturally potent versus the difference of mere caretaking as measured against original creativity; and the assertion of the life of the Louvre as the scene of art in France opposed to the death that haunts museums as mausoleums of past genius. Pei's project, then, although this does not at all imply conscious intention on his part, is a strategy to assert and attain intelligibility and power. The pyramid is not so much a formal, aesthetic object or worthy museum piece as an act.

Of course, the act of Pei's pyramids simultaneously dislocates earlier French uses of pyramids. The pyramid of the Mitterrand government asserts itself as a monument of great individuals (Mitterrand, Pei, and others) and current French culture, to be admired and emulated, and also as conferring a sort of cultural immortality on them. Nevertheless, that posture shows, for instance, that the eighteenth-century neoclassic architectural memorials designed by Boullée, Dufourny, and others[32] fail to sustain their supposed earlier accomplishment. Their meaning as access to sublime nature and thus timeless divinity is no longer operative. Their testimony to the triumph of almost forgotten individuals over death is not credible. If the neoclassic vision of architecture and death really had the meaning it purported to have, it would still be powerful. Obviously, it is not.

Therefore, repetition is required. As the French earlier had appropriated the antique and classical to assert their presence and identity, now the ritual must be repeated. As Louis XIV, for example, asserted his and France's presence, identity, and life in rebuilding the Louvre's Petite Galerie (continuing the process of reconstructing the Louvre begun by Charles V in the thirteenth century); as Napoleon asserted it in his day with power over past empires and the Egyptian pyramids and the Sphinx themselves; as Boullée, Jacques Rousseau, Blondel, and Ledoux had at the end of the eighteenth century with their architecture of death; so do Mitterrand and Pei today.

In asserting cultural presence and their immortality as patrons and creators of art, the Mitterrand government and Pei confirm the conception of immortality as "lasting artistic and cultural reputation" and again ground the Louvre as monumental museum, where the new work joins the display and commemoration of earlier works, artists, and patrons. Here is the completion of a turn begun by French neoclassic architecture. Originally, for the Egyptians and orthodox Christianity alike, immortality meant participation in the timeless realm of divinity. Later, with Boullée and his colleagues, it partially indicated that same earlier meaning through the sublimity of nature, but the concept also diverged, since in affirming inspiration to glorious deeds, the eighteenth-century French finally understood timelessness as commemoration that lasts in citizens' minds. It seems natural that, as the belief in divinity gave way to faith in rationality, *immortality* would come to mean "remaining everlasting, as present in human consciousness."

The force and meaning of Pei's pyramids therefore give no hint of support for lost and bankrupt ideas such as eternal presence with divinity. In this regard the Louvre is the perfect site for Pei's pyramids. Earlier the scene of assertion of the Sun King's power (Louis XIV identified himself with the sun as symbol of the presence of divine right in the tradition of the pharaoh empowered by his sun god, Ra), now it is the site for the claim of current power and identity that is accomplished by building over, and thereby incorporating, tradition.

Needless to say, the actions made possible by Pei's act of designing the pyramid are more mundane than those in the Egyptian (or neoclassic) era. Pei's pyramid provides ground-level access to the facilities, and whereas the Egyptian pyramids hid the pharaoh's body to allow for his "underground" eternal life, Pei's pyramid (or really, the public reaction to and focus on it, which distracts us) conceals Pei's larger architectural task: expanding the underground parking facilities and promoting tourism. As Derrida would have it, it is a case of the pit reasserted over the pyramid.[33]

Architecturally and epistemologically, presence and immortality are again affirmed as dominant over absence and death, but now this assertion occurs through new concepts covertly usurping the definition of the original ones; that is, the mastery depends on the inversion of the traditional meanings. Originally presence and immortality meant the dominance of deity over beings, of life in the otherworld over life in this world, of the unbroken succession of timeless meanings and forms in the built environment over constant change and idiosyncratic variation. Understood as the displacement of the earlier traditional architecture of eternal presence by continuous reaffirmation of temporal presence, Pei's pyramids have their meaning as part of our posture and strategy for today's power. We have, then, a way to interpret Pei's gloss on his pyramid, as noted by an architectural reporter: "He described it, rather enigmatically, as having an architectural presence while being less than architecture."[34]

Finally, Arata Isozaki's new Museum of Contemporary Art (MOCA) near downtown Los Angeles also appropriates the Egyptian pyramids. In the final scheme for the red sandstone project, pyramidal skylights are positioned over some of the galleries. On one wing there are two rows of four pyramidal skylights; on the other, three pyramids, two small ones (the same size as the eight on the other wing) before a large one, evoking Giza (see fig. 2.11). Together with the other disparate building elements (notably, a barrel-vaulted entrance and sunken court), the museum as a whole is an interesting candidate for interpretation according to the reigning architectural theories. For example, one largely sympathetic review sees the pyramids and building forms practically (technically and prudently) and judges that they are a decent solution to arbitrary and poor site and development restrictions. Similarly, the museum's

Figure 2.11. Pyramidal skylights over galleries. Arata Isozaki, Museum of Contemporary Art, Los Angeles, 1981–1986.

six successive schemes are seen through Venturi's influential *Complexity and Contradiction in Architecture* as developing from an initial attitude of repose toward contradiction and a postmodern style.[35]

It is more than ironic that Isozaki introduces pyramids into the design at the point where it moves out of repose and into contradiction: Egyptian sacred form and repose displaced, indeed. Not surprisingly, the notions of contradiction and repose do not enable an adequate interpretation of the pyramids, much less of the palm trees originally intended to line the street outside. For example, the previously cited critic wonders whether the juxtaposition of pyramid and palm trees is a joke.[36] Of course, postmodern architecture is full of wit, but there is more going on here.

In one way the project shows the dissolution of the pyramid and architectural form, from reflecting culturally anchored and validated dimensions with generative and normative power to participating in a stock of "equally meaningful" and "homogeneously valued" design resources.[37] Los Angeles is the paradigmatic site of such understanding. The city has its character and charm precisely as the locus for the casual and humorous, albeit sophisticated, blending of exotic forms and styles, especially "tropical" ones. The Museum of Contemporary Art's pyramids are right at home with the city's tradition of "Egyptian" restaurants, theaters, and stores.[38] Thus, Isozaki indeed has "gone native" (a comment made, but not developed, by a reviewer). It can be said that just this contemporary attitude, by its humor and casual enjoyment, makes a comfortable place for our presence and identity by dislocating what appear as pompous and stuffy attitudes and architectural forms of high culture (e.g., the Louvre). With its pyramids MOCA is just right to house art in Los Angeles and assert power and the purpose of art today.

Furthermore, by presenting the sandstone building, pyramids, and palm trees together, the project does not so much allude to the Egyptian desert and the French tradition of cemetery interpreted as Egyptian desert (surely ·the latter connection would not at all have been intended by Isozaki) as "give them the palm." The Egyptian pyramids, strikingly preserved even while being worn away and covered by the desert sand, no longer display the powerful life of a civilization formed to be an oasis of eternal life. In the French neoclassic adaptations the pyramid is already an Egyptian wasteland bespeaking only death; according to the culture of the sublime, life is affirmed only in nature's fertile but impersonal timelessness. Today the pyramids are appropriated by MOCA, which celebrates Los Angeles as a present, vibrant, and extravagant "desert culture" and as now being what the Egyptian world once was but is no more.

A final displacement. The Museum of Contemporary Art's pyramids are skylights that, copper glazed, will reflect their surroundings. Isozaki's and Pei's pyramids are the same, then, in dealing with light as a natural phenomenon: these pyramids let in daylight so that we can see, let light shine out at night, and, mirroring, reflect the city and sky around them. Pei's and Isozaki's postmodern pyramids, in their relation to light and life, are altogether opposite to the traditional Egyptian (and neoclassic) pyramids (see fig. 2.12).

Recall that the pyramids of Egypt gathered and held the divine force of life, which emanated from the sun god, Ra, to his son, the pharaoh, and thus to the kingdom, as did the rays of the sun.

Figure 2.12. Museum of Contemporary Art's pyramids deal with natural light. Arata Isozaki, Museum of Contemporary Art, Los Angeles, 1981–1986.

Accordingly, light had cosmic significance for the Egyptians. Of course, ordinary, profane sunlight also made people sweat and crops grow, but the soul lived, plants grew, and people saw—that is, light was efficacious—only because it first of all was sacred power. The pyramids focused and reflected such sacred light from their gold capstones and closed out profane, ordinary light from their permanently darkened interior passages and chambers. Inside the sealed pyramid the pharaoh's *ka* remained in union with the sun god and moved about, continuing his action in his eternal dwelling.[39]

In a way opposite to this conception, Pei's pyramid at the Louvre and Isozaki's at MOCA deal with light as a fully natural phenomenon. What else would it be for contemporary architecture? Light is modulated and dispersed for use and treated as a design element because of its properties in regard to shadow and range of saturation. As the museums' pyramids allow light to enter into the buildings, as they reflect the cities' skies, and as they let light shine out, they extinguish the meaning of the Egyptian pyramids and their supposed ground in sacred light.

As noted, the pharaoh's pyramid kept natural light out and opened a site within for divinely emanated light, which enabled the pharaoh's *ka* to live in the power of a richly symbolic world, housed in chambers full of beautiful objects whose sacred forms and meaning had been given by the gods. Pei's and Isozaki's pyramids let in humanly controlled sunlight so that we can enjoy the colorful objects we have made and collected for ourselves.

The Egyptian pyramids symbolically reflected the vital force that the gold capstones physically reflected; that is, they showed not the pyramids' profane surroundings but the sacred cosmos. The pyramids in Paris and Los Angeles reflect the surrounding human and natural environment, the city's lights and passing weather. No physical light emanated from the Egyptian pyramids, but because the pharaoh lived eternally within, he continued to disperse to the kingdom the vital force that he alone possessed as the gift from the sun god. Pei's pyramidal "lantern," in contrast, shows our power. Even when the sun is gone, we generate light to illuminate what we have created and thus provide for our own culture and way of life.

The new pyramids also have reversed the meaning of life. The pyramid was originally the scene of life and presence because it was the place of sacred life, reserved for the gods and their sons, the pharaohs. Accordingly it was located in the desert within a necropolis, a city of the dead (those with eternal life in earthly death), physically separate from and inaccessible to ordinary communal life. Today, however, death and eternal life in an otherworld are banished by the processes of earthly life. Gods and pharaohs are replaced by participants in democratic societies. The pyramids help to establish our secular presence and identity and belong within the city of the living.

In the end the displacement of the earlier meanings of the same forms by these postmodern inversions accomplishes for us what earlier realms attempted to do in their time. Although the succession of forms and meanings shows the futility of expecting final success, much less the experience of any eternal presence, the pyramids provide a tactical device by which we can maneuver, trying to establish and maintain our identity and presence as part of a city of the living. Where better to celebrate the triumph and limitations of our lives and the power, posture, and wit of the free play of our art and architecture than in the center of the Cour de Napoléon or downtown Los Angeles? How better than with displaced pyramids?

Chapter **3**

Hermeneutic Retrieval

American Nature as Paradise

3
Hermeneutic Retrieval
American Nature as Paradise

America Religiously Understood

The starting point of this chapter is a general claim: originally the American understanding of nature was substantially religious, which means that attendant practices operated according to a theologically informed economy and politics.[1] What we now take to be matters of science (for example, the wilderness as interpreted by geology or physical geography and, perhaps, by materialist cultural geography, insofar as nature is modified by humans) originally were understood by means of what historian of religion Mircea Eliade calls "mythical geography."[2] Specifically, from the beginning, America was understood in terms of an earthly paradise.[3] This tradition has its source in Genesis, where the *second* creation story (2:8–17) describes the Garden of Eden.

> And the Lord God planted a garden eastward in E'den; and there he put the man whom he had formed. And out of the ground made the Lord God to grow every tree that is pleasant to the sight, and good for food; the tree of life also in the midst of the garden, and the tree of knowledge of good and evil. And a river went out of E'den to water the garden; and from thence it was parted, and became into four heads. (Gen. 2:8–10)

> And the Lord God took the man, and put him into the garden of E'den to dress it and keep it. (Gen. 2:15)

The interpretation of America by way of Scripture was fundamental: Columbus, according to his *Book of Prophecies,* believed that he had discovered the Garden of Eden and thus had enabled a significant advance in the conversion of the world and its consequent end. He wrote, "God made me the messenger of the new heaven and the new earth of which he spoke in the apocalypse of Saint John, . . . and He showed me the spot to find it."[4]

This was not an idiosyncratic attitude toward the New World, regardless of how accurately it explains Columbus's place in the scheme of events. Throughout the fifteenth and sixteenth century people believed that the time had come to renew the Christian world and that this renewal was to be the return to the earthly paradise or the beginning of a new era of sacred history.[5] That is, America was to be the scene where the Church would complete its work and Christ's second coming would occur.[6]

Proceeding from the English Reformation, the colonization of America elaborated such a sacred history, which progressed in the movement west; the unfolding spiritual drama was seen as fulfilling the typology of America's mission.[7] Thus, in the seventeenth century it was common to think of the sun's course in terms of a spiritual journey, so that one could follow the path to paradise in the west (for example, as Thomas Burnet and Bishop Berkeley did).[8] Ulrich Hugwald prophesied that in this new era humanity would return "to Christ, to Nature, to Paradise."[9]

The early maps of America during this period concretely located the understanding of the anticipated Garden of Eden: Guillaume Le Testu's pen, ink, and watercolor map *Terre de la Floride,* which appeared in his manuscript atlas *Cosmographie Universelle* (1555), shows the east coast as parkland with game (see fig. 3.1). And *Historiae Canadensis* (1664) depicts Indians easily gathering seabirds, reinforcing the view that America was populated with partridges too big to fly and turkeys as fat as lambs.[10] Indeed, from early descriptions by Giovanni da Verrazzano, William Wood, and others, it appears that along the southern coast of New England, from the Saco River in Maine all the way to the Hudson, the woods were

Figure 3.1. Guillaume Le Testu, *Terre de la Floride,* from the manuscript atlas *Cosmographie universelle,* 1555. Manuscript, pen and ink and watercolor, 35 cm x 48 cm. Ministère de la Défense—Service Historique de l'Armée de Terre, Vincennes.

remarkably open, almost parklike at times. In 1614 John Smith spoke of New England as an Eden: "heaven and earth never agreed better to frame a place for man's habitation . . . we chanced on a lande, even as God made it."[11] This attitude still appeared in 1737 with the illustration of William Byrd's proposed townships entitled *Eden in Virginia.*

As a topic, however, an interpretation of America aligned with Christian thought is vast and may seem temporally distant. Even granted that we can start our investigation with an America understood religiously from its very beginning, we still have the long passage to today, where such an idea seems quaint at best, if not totally anachronistic. To focus our study and concentrate on the transition, we can ask how such a religious tradition was interpreted and used by nineteenth-century painters to elaborate a paradigmatic understanding of American nature and, hence, the landscape (an original understanding that has changed into that which prevails today).

Two main lines of development will concern us here, both derived from Genesis and specifically from its two creation accounts, two traditions that are unified in the source from which they spring.[12]

A Natural Paradise Already Given

The initial problem in the American understanding and evaluation of the landscape was that it did not appear as landscape. That is, it did not qualify as landscape according to the reigning European conventions of the cultivated-natural, which were associated with previous periods of civilization. Because America was so unlike Europe and without a history, it was difficult not merely to paint but even to perceive.

The problem and difficulties may be clearer if we make a leap to consider the solution first. In the development of the religious tradition of nature as paradise, America becomes an agrarian paradise. Here, the farmer is likened to the new Adam. George Washington writes to Lafayette: "Americans should be employed in the . . . agreeable amusement of fulfilling the first and great commandment—Increase and Multiply: as the encouragement to which we have opened the fertile plains of the Ohio to the poor, the needy

Figure 3.2. William T. Ranney, *Daniel Boone's First View of Kentucky, 1849.* Oil on canvas, 36" x 53½". Courtesy of the Anschutz Collection, Denver.

and the oppressed of the Earth; anyone . . . may repair thither and abound, as in the land of promise, with milk and honey."[13]

The image is captured in William Ranney's painting *Daniel Boone's First View of Kentucky* (1849; see fig. 3.2), which was based on a passage in Timothy Flint's popular biography of Boone:

> They stood on the summit of Cumberland mountain. What a scene opened before them! A feeling of the sublime is inspired in every bosom susceptible of it, by a view from any point. . . . They remarked with astonishment the tall, straight trees, shading the exuberant soil, wholly clear from any other underbrush than the rich canebrakes, the image of verdure and luxuriance, or tall grass and clover. . . . This wilderness blossoms as the rose, and these desolate places are as the garden of God.[14]

From a stagelike promontory Ranney's figures gesture toward the "promised land" and appear entranced with dreams of the future, an effect underscored compositionally by the movement from right to left, toward the west, source of the warm, beckoning light.

The question becomes, then, how did this stereotype emerge? How did the interpretation of an agrarian paradise get worked out in nineteenth-century landscape painting?

We can follow the growing recognition beginning in the 1820s by attending to the works of Thomas Cole, who was among the first fully to delineate the wilderness as a desirable subject for painting; that is, Cole understood wilderness as a subject matter endowed with religious and moral significance and hence as a profound symbol of the new nation.[15] The entire course of Cole's work developed this interpretive understanding.

It is not surprising that the assumptions in Cole's early work are largely European. For example, in his *Italian Scenery* (1833) we see the result of the idea that to be a landscape at all, nature must be cultivated: landscape is humanly improved nature, as modified and imagined according to classical conventions. The terrain in this painting is laden with the heritage of Western culture: ruins bespeaking the ancient past are represented by the aqueduct and fortification in the middle ground and temple in foreground; there is the picturesque view of trees, lake, hills, mountain, and pacific sky derived from modern attitudes toward scenery and "view"; and the entire scene evokes the complex pastoral tradition of celebration and meditation in the humanized landscape steeped in religious memory.

Another, related set of European conventions also plays a part in Cole's background, namely, those conventions bound up with the tradition of the sublime, in which nature is so overwhelming, so astonishing, that the individual cannot but see and reflect on God. A classic example is Turner's painting *Valley of Aosta: Snowstorm, Avalanche, and Inundation* (1836), which depicts the three awe-inspiring phenomena occurring simultaneously. Because God created the earth, for us to see it in its full grandeur is to lose ourselves and to catch a glimpse of him. The sublime, then, draws from the lesson of the Book of Job, which climaxes, after Elihu's hymn to God ("He spreads out the mist, wrapping it about him, and covers the tops of the mountains. He gathers up the lightning in his hands"

Figure 3.3. Thomas Cole, *Landscape with Tree Trunks, 1828*. Oil on canvas, 26½" x 32½". Museum of Art, Rhode Island School of Design, Walter H. Kimball Fund.

[Job 36:31–32]), with the description of theophany, evoking the power of the Almighty, before whom Job finally bows down.

As early as 1828, in *Landscape with Tree Trunks* (see fig. 3.3), Cole endows the American landscape with features of the sublime: the mountains, clouds, trees, and water evoke the great and sublime in nature and open the spectator to a contemplation of the deity. This is clear from at least four crucial elements: the distant light contrasts with the passing storm, embodying or manifesting the aesthetics of the sublime; the monolithic mountain appears as a focal symbol of theological monotheism; anthropomorphic trees heighten the drama; a cross, with its deliberate Christian associations, appears at the tip of a tree. What is amazing is that although Cole understandably relies on European conceptual, formal, and compositional conventions in this painting, he nonetheless takes a major step beyond them. He depicts a scene that is a distinctly American wilderness and endows it with moral and religious symbolism.

Thus, in this painting we see Cole's first key move: to understand American nature by way of the Christian interpretation of nature as creation and, as a lasting symbol and manifestation of God. Cole himself uses this interpretation in his *Essay on American Scenery:*

> But in gazing at the pure creations of the Almighty, [one] feels a calm religious tone steal through his mind. . . . There are those who regret that with the improvements of cultivation the sublimity of the wilderness should pass away: for those scenes of solitude from which the hand of nature has never been lifted, affect the mind with a more deep toned emotion than aught which the hands of man has touched. Amid them the consequent associations are of God the creator— they are his undefiled works, and the mind is cast into contemplation of eternal things. . . . Look at the heavens when the thunder shower has passed, and the sun stoops below the western mountain—then the low purple clouds hang in festoons around the steeps—in the higher heaven are crimson bands interwoven with feathers of gold, fur for the wings of angels; and still above is spread that interminable field of ether whose color is too beautiful to have a name.[16]

In a masterful synthesis, then, Cole combined the sublime view of nature with a specific religious and historical vision, as is further detailed in his writings, especially the poem "The Cross," his letters, and the *Essay on American Scenery* (1835), as well as in a lifetime of paintings with explicitly Christian subject matter, such as the series of five monumental paintings (each 64 in. x 96 in.) entitled *The Cross and the World* (1846–1847; now lost). From the remaining studies for that series, such as *The Vision: Study for Cross and World* and *Pilgrim of the Cross at the End of his Journey* (1847), we learn beyond doubt that, for Cole, light and the cross (as well as cloud-angels) were embodiments of God's glory and thus revealed his divinity in the landscape.

Even more to the point, perhaps, is to compare *Landscape with Tree Trunks* with another work painted at the same time, *Expulsion from the Garden of Eden* (1827–1828). The striking and unusual anthropomorphic, storm-blasted trees and the mountains with associations to divinity are common to both. (Not incidentally, the latter work belongs with a recently rediscovered companion painting, *The Garden of Eden.*)

More important, although they are not formally a pair, these works show two of the crucial stages of the story closest to the Christian mind in America: after the Creation and mediated by the Crucifixion and Redemption, we have the expulsion from the Garden of Eden and the promised second coming of paradise. Thus, where brute wilderness would be the fallen state of nature, the punishment after Original Sin and the expulsion from the garden, we see as more striking its contrast with American nature portrayed as blessed wilderness, that is, where wilderness boldly is interpreted as paradise regained.

Cole moves another step further when he inserts a human into the scene of natural wilderness in *Daniel Boone and His Cabin at Great Osage Lake* (1825–1826; see fig. 3.4). His portrait of Daniel Boone reveals the hunter juxtaposed before the wilderness, the latter such an impenetrable tangle as to preclude settlement, making clear Cole's belief that humankind should not exploit the wilderness. As confirmation of the vision of wilderness as a natural place excluding Europeans, and Cole's deliberation concerning any move beyond that state, it is important to note the dramatic change in the final version of this painting: in a preliminary drawing for the work the figure was not Boone but a Native American, the indigenous inhabitant.

The final shift, or really the radical completion and disclosure of the underlying and unifying seminal understanding, was finished twenty years later in *Home in the Woods* (1847; see fig. 3.5). In this painting we discover a sanctified image of rural life, with a confirmation of the blessed life of the pioneer—if such a person is virtuous in this abundant land.[17] There is little evidence of agricul-

Figure 3.4. Thomas Cole, *Daniel Boone at His Cabin at the Great Osage Lake, 1825–1826.* Oil on canvas, 38" x 42½". Mead Art Museum, Amherst College, Massachusetts. Museum Purchase.

ture or crops because humankind is not condemned to toil and sweat in a fallen state. Rather, the pioneer in a new Eden lives effortlessly from the bounty of the wilderness, without work or hardship. In the painting we see his return from fishing, greeted by his family before an amazingly prosperous cabin. Further, the iconography makes clear a religious meaning: the trellis/cross on the side of the house denotes the Christian life of the pioneers, as do the three trees behind the cabin and the monolithic mountains in the distance.[18] Over the pregnant scene, a clear sky presides, of which Cole wrote, "the pure blue sky is the highest sublime. There is the illim-

Figure 3.5. Thomas Cole, *Home in the Woods, 1847.* Oil on canvas, 44" x 66". Reynolda House Museum of American Art, Winston-Salem, North Carolina.

itable. . . . There we look to the uncurtained, solemn serene—into the eternal, the infinite—toward the almighty."[19]

Cole justifies the final form of the vision held in *Home in the Woods* as follows: "I have alluded to wild and uncultivated scenery, but the cultivated must not be forgotten, oft it is still more important to man in his social capacity . . . , it encompasses our homes, and, though devoid of the stern sublimity of the wild, its quieter spirit tenderly into our bosoms intermingles with a thousand domestic affections and heart touching associations—human hands have wrought, and human deed hallowed all around."[20] Again, and crucially, *Home in the Woods* shows not toil and exploitation but harmony and natural bounty in the land given to Americans.

Cole's development of the wilderness theme from 1827 to 1847 works out the interpretation of the American landscape first as a gloriously pristine natural paradise, next as the natural site contain-

Figure 3.6. Frederic Edwin Church, *To the Memory of Cole, 1848.* Oil on canvas, 32" x 49½". Des Moines Women's Club—Hoyt Sherman Place, Des Moines, Iowa.

ing humans, and finally as the natural dwelling place that shows itself as the inhabited Garden of Eden. The final version in no way lessens or replaces the earlier two; indeed, it completes them and brings forth the sustaining vision, that is, the religious understanding underlying the problem and solution. How, then, does the interpretation of America develop? *As a gradual meditation on the second creation account in Genesis.* Cole succeeds in a delineation that belongs with the earlier religious understanding of American nature as paradise.

That both the mantle of Cole's prestige and the importance of Christianity's historical and allegorical interpretations of the American landscape were of central importance to contemporary artists is confirmed by a crucial painting by Cole's student Frederic Edwin Church, only recently relocated. In 1848, immediately after Cole's unexpected death, Church produced *To the Memory of Cole,* a bold work that fuses overt symbolism and nature (see fig. 3.6).[21] In the foreground of the painting a cross festooned with roses and vines "re-presents" the central image of Cole's *Cross in the Wilderness*

Figure 3.7. Frederic Edwin Church, *New England Scenery, 1851.* Oil on canvas, 36" x 53". George Walter Vincent Smith Collection, George Walter Vincent Smith Art Museum, Springfield, Massachusetts.

(1845). Church identifies with Cole both by appropriating the cross motif and—in a move that already is somewhat retrograde in the light of the naturalistic trends of midnineteenth century—by insistently making explicit what was quietly present in Cole's *Home in the Woods.* (Church reaffirms the shared religious understanding in *Apotheosis to Thomas Cole* [c. 1868], in which a cross appears in the sky to beckon the two figures as pilgrims on a holy path.)

Although the gifted Church obviously wanted to acknowledge and operate within the sphere that Cole had delineated for landscape painting, he also transformed what he "inherited," participating in the development of naturalistic images that yet remained deeply moral and heuristic.[22]

In another early painting, *New England Scenery* (1851), also painted under the influence of Cole's example, Church transposes Cole's pioneer home in the wilderness to a later phase of the American inhabitation of nature. Whereas Cole's *Home in the Woods* might be conceived of as representing a state of primitive arcadian bliss, Church's *New England Scenery* represents a fully developed agricultural paradise, the core of a soon-to-be consummated civilization.

New England Scenery is illuminated by the warm glow of afternoon light (see fig. 3.7). In this study Church rehearses all the compositional elements of the Claudian tradition, transforming them into characteristically North American wagons, cabins, mills, villages, and attendant domestic animals. While the eye is absorbed in detailed vignettes of tranquil and comfortable country life, a covered wagon heads west to develop the frontier. Patches of cleared land are bounded by the vast potential of the yet-unclaimed wilderness; the wagon is beckoned by a series of openings through woods and mountain valleys, presided over by majestic trees, mountains, and clouds. The entire scene has a tone of ease and welcome. The youthful Church has imaginatively synthesized all the characteristic features of the agrarian myth and European tradition into an ideal landscape of extensive proportions.

Church's masterful *Twilight in the Wilderness* (1860) echoes Cole's vision—a settler's cabin was in a study but omitted from the final work—even as it clearly moves beyond it. Here we have a major step, deepening the vision through gradually eliminating references to civilization and creating a uniquely American image of wild nature. That is, once the fundamental interpretation of American nature as paradise for inhabitation is established, the European conventions can be transcended altogether and thus eliminated. In addition, once it is understood that the wild is the American Garden of Eden, it can be painted more subtly and indirectly, without explicitly insisting on human figures. With the new mythology in place, the explicit narrative can, if desired, be omitted in favor of a more "symbolic" treatment. Crucially, the interpretation can be presented in a seemingly naturalistic fashion precisely because the perception of this revelation of nature as paradise was self-evident to those initiated at the time and thus internalized. It is able to disappear as the "unspoken."

In *Twilight in the Wilderness* Church transformed the conventional stereotypes of wilderness painting and created an original

Figure 3.8. Frederic Edwin Church, American, 1826–1900. *Twilight in the Wilderness, 1860.* Oil on canvas, 40" x 64". The Cleveland Museum of Art, Mr. and Mrs. William H. Marlatt Fund, 65.223.

Figure 3.9. Frederic Edwin Church, *Heart of the Andes, 1859.* Oil on canvas, 66⅛" x 119¼". The Metropolitan Museum of Art, bequest of Margaret E. Dows, 1909.

image of the New World's archetypal wilderness through his delineation of light, clouds, mountains, water, and trees (see fig. 3.8). Here only the primeval experience of nature endures. The sky's seething reds and cool blues and the chorus of angel-like clouds announce the glory of God's presence for the nineteenth-century Christian spectator. The awesome truth of nature's divinity in America is manifested, renewing the promise of the New World that God had offered as America.[23]

Church further transforms this fervent vision, now in a nationalistic direction, in *Our Banner in the Sky* (1860). Here the vision is expanded dramatically, beyond the frontier to the pathless wilderness. We find no sign of human penetration, only a single bird, even as a divine light blesses the entire landscape in the form of a U.S. flag glowing in the heavens at dawn.

Church's definitive image of natural divinity in *Twilight in the Wilderness* was complemented by South American landscapes, which also were interpreted in terms of religious vision and which made clear that both continents comprised the New World as natural paradise. In Church's *Heart of the Andes* (1859; see fig. 3.9), which combines the divine and the natural, we find a keynote in the white wayfarer cross—again the cross in the wilderness—that focuses the meaning for us: paradise regained through the mediation of the Redemption. Eden is promised as again possible, even actual. As contemporary reviews make clear, viewers saw the painting as depicting a "Paradise" and an "Eden." Moreover, in 1862 Church rendered *Cotopaxi,* a primal earthscape that clearly is geography viewed through Genesis. The earth freshly appears before us; the newly risen sun discloses an edenic jungle in this image of cosmic history written in the landscape.

Lest it be thought that depicting the American landscape as the manifestation of the divine is merely idiosyncratic to Cole and his admiring disciple Church, and not the definitive articulation of a culturally shared vision, consider one of Church's rivals, Albert Bierstadt. Indeed, such was the impact of Church's work that Bierstadt, once thought to be largely indifferent to religious matters, nonetheless attempted to emulate Church's *Twilight in the Wilderness.* This is a decisive point for the diffusion of the vision, since the work of both Bierstadt and Church was enormously

Figure 3.10. Albert Bierstadt, *Sunset in Yosemite Valley, 1868.*
Oil on canvas, 35½" x 51½". Haggin Collection, The Haggin Museum,
Stockton, California.

influential through popular exhibitions and widely distributed etchings based on their major paintings.

A series of Bierstadt's works indicates the development of the theme. The culminating work, *Sunset in Yosemite Valley* (1868), derives its power from the astonishing contrast of light and darkness. On the left mountain towers rise like cathedrals; on the right we find El Capitan, whose shadow intersects the river below, dramatically forming a cross on the valley floor, a cross in and of the landscape. The scene is presided over by clouds above El Capitan formed into angel shapes (see fig. 3.10). God's sublime power emanates from the sky as majestic light presiding over a decidedly paradisiacal valley. The connection to the second creation account, clearly conveyed by the intensity of the sunset and the awesome golden-toned reflection in the clouds, was all too obvious to contemporary viewers. As one writer put it, the work shows the landscape as a "great natural temple of sublimity." Then too, Yosemite itself often was seen as an example of a natural paradise. Bierstadt

Figure 3.11. Albert Bierstadt, *The Oregon Trail, 1869.* Oil on canvas, 31" x 49". The Butler Institute of American Art, Youngstown, Ohio.

first visited Yosemite in 1863 with the writer Fritz Hugo Ludlow, whose account, which appeared in *Atlantic Monthly* and later in a book, makes the interpretation explicit. Of their expectations, for example, he says, "If report was true, we were going to the original site of the Garden of Eden."[24]

Bierstadt builds on this dramatic work in the quieter, but no less glorious, second version of *The Oregon Trail* (1869), which assumes the accomplishment of *Sunset in Yosemite Valley* (and thus of Church's *Twilight in the Wilderness*) and symbolically moves within the shared understanding. In *The Oregon Trail* the landscape again opens for human life (see fig. 3.11). In the foreground we find settlers entering the blessed land in an image of economic order. At right, in the middle distance, settlers peacefully share the spacious valley with Native Americans, while in the far distance the majestic goal beckons. The whole is bathed in warm, golden light and promises easy movement along the dominant diagonal compositional axis. The human place in the garden is granted and held open by divine grace; the dramatic, palpable quality of light, especially the rays that halo the sun, expresses the belief in God's approval by means of the contemporary convention.

In short, wilderness is shown to be both the uniquely American characteristic and the bearer of meaning since it is given as an earthly paradise, that is, as a physical and moral dwelling place. Furthermore, the articulation of nature as paradise is worked out and cultivated across a range of images, with, at one pole, the symbolic, "purely natural" wilderness infused with the significance of the divine and, at the other pole, the explicitly edenic garden landscape humanly inhabited.

A final phase testifying to the widespread acceptance of the vision of divinity in the landscape is exemplified in an event, symbolically prepared for and almost anticipated, that consummates the vision. The dramatic discovery of the Mount of the Holy Cross, that is, of a cross of snow and ice on the face of a mountain, was made by explorer Ferdinand V. Hayden's expedition to western Colorado in 1873, whose company included photographer William H. Jackson. Jackson's images of the mountain are among the classic images of the American West, and his photograph of the phenomenon was distinguished with a medal in the 1876 Centennial Art Exhibition in Philadelphia.

Here, no longer generated only in the vision of artistic imagination for an initiated audience but naturally presented for the perception of all, both those inclined to see landscape through religious interpretation and those not so disposed, the American landscape itself witnesses the same fusion: American nature and symbol of divine grace. The phenomenon was celebrated by Thomas Moran in his famous painting *The Mountain of the Holy Cross* (1875; see fig. 3.12) and then by Longfellow, who hung a print of the painting beside a portrait of his deceased wife and wrote the poem "The Cross of Snow."

An article in the *Illustrated Christian Weekly* of 1875 described Moran's painting: "Suddenly the artist glances upward, and beholds a vision exceedingly dramatic and beautiful. He is amazed, he is transfixed. There set in the dark rock, held high among the clouds, he beholds the long straight cross, perfect, spotless, white, grand in dimension, at once the sublimest thing in nature and the emblem of heaven."[25]

Figure 3.12. Thomas Moran, *The Mountain of the Holy Cross, 1875.*
Oil on canvas, 82¾" x 64¾". Gene Autry Western Heritage Museum,
Los Angeles.

At this point we find the climax of understanding American nature in terms of Genesis's second creation account. On the one hand, the interpretation is widely known and accepted; on the other, it is about to wane. Indeed, what follows is not so much anticlimatic as almost silent, so quietly and quickly does the once daring and powerful vision fade. The rise of secular materialism and science, dominant in the last quarter of the nineteenth century, helped to end this religious tradition.

The work of geologist Clarence King in Yosemite, for example, displaced any but the secular materialist, scientific manner of seeing the natural environment. King held that nature does not bear any such meaning as the previously mentioned painters believed, contending that such interpretations need to be cleared away in favor of seeing the physical earth as a product of "evolutionary" forces. King, who certainly was not without a sense of wonder and beauty, does more than capture differences in climate and atmospheric effect when he contrasts the view of Yosemite's El Capitan on October 5, 1864, to the same view in June:

> Now all that [sublimity] has gone. The shattered fronts of walls stand out sharp and terrible, sweeping down in broken crag and cliff to a valley whereon the shadow of autumnal death has left its solemnity. There is no longer an air of beauty. In this cold, naked strength, one has crowded on him the geological record of mountain work, of granite plateau suddenly rent asunder, of the slow, imperfect manner in which Nature has vainly striven to smooth her rough work and bury the ruins with thousands of years accumulation of soil and debris.[26]

King quite consciously led the life of the scientific individual; even the formal structure of his writings rhetorically moves the reader to accept the scientific view since, although occasionally allowing himself to lapse into an imaginative sympathy with "mythic" and aesthetic perceptions, in the end King regains the "saving" clarity of the scientist. In his judgment the archaic thought or mythmaking of primitive peoples and many artists (which, he believes, still "smoulders in all of us"), and the attendant "burden of a hundred dark and gloomy superstitions," obscures natural, material reality: "The vary-ing hues which mood and emotion forever pass before his own men-tal vision mask with their illusive mystery the simple realities of nature, until mountains and their bold, natural facts are lost behind the cloudy poetry of the [artist]." In contrast, King champions modern scientific thought, for example, in "realizing fully the geo-logical history and hard, materialistic reality of Mount Whitney, its mineral nature, its chemistry": "as the [symbolic gaunt, gray old Indian] trudged away . . . I could but feel the liberating power of modern culture, which unfetters us from the more than iron bands of self-made myths. My mood vanished with the savage, and I saw the great peak only as it really is—a splendid mass of granite 14,887 feet high, ice-chiselled and storm-tinted; a great monolith left standing amid the ruins of a bygone geological empire."[27] By the end of the century, then, as a public view, the landscape of divinity and a great part of midcentury understanding of natural paradise was replaced.

Paradise Promised: Wilderness to Be Converted

Of course, the nineteenth-century religious interpretation of the landscape was more complex than the preceding account indicates. Indeed, the story is incomplete without recognition of the dynamic generated by the tension between the interpretation of nature as paradise and an alternative, more powerful interpretation drawn from the same religious and biblical tradition. In dramatic contrast to the interpretation of American wilderness and the subsequently inhabited landscape as a natural Garden of Eden was the even more commonly understood, well-documented interpretation of American wilderness as opposite to paradise.[28]

Here, although America also was understood as the promised site of a second paradise, paradise was held to be wild nature cultivated and subjugated. That is, in the alternative view, natural wilderness is a wasteland or desert, and only the transformation and conquest of that barrenness can generate the second paradise, or heaven on earth. This interpretation derives from the other, culturally domi-nant account of creation, which does not speak of paradise and which appears first in Genesis.

In this account, after creating the heavens and the earth, God specifically elaborates the place and role of humans in, but distinct from, the "natural" world. As the King James version of the Bible renders it, "And God blessed them, and God said unto them, 'Be fruitful, and multiply, and replenish the earth, and subdue it: and have dominion over the fish of the sea, and over the fowl of the air, and over every living thing that moveth upon the earth'" (Gen. 1:28).[29]

The early settlers took this text most seriously in their understanding of America as a desert to be overcome and, like the Old Testament's exodus from Egypt, as a trial before passage to the promised land, the promised paradise on earth.[30] The Puritans, for example, saw the American wilderness in the light of the expulsion from paradise. Hence, against the Old and New Testament background, William Bradford, embarking from the *Mayflower*, encountered what he described as a "hideous and desolate wilderness."

As a consequence American nature came to be seen in relation to the idea of work. As humans were expelled from paradise and given the religious task of recovering, with the aid of Christ, a place before God, so too the new "paradise on earth" would be produced by work. The destruction of the wilderness was the first step toward building the new kingdom, an idea that is elaborated by Jonathan Edwards in the first half of the eighteenth century. In short, the task in both its religious and civic dimensions was to control and cultivate the wild and to make nature into paradise in America. Although the sublime signifies God's blessing, the interpretation from Gen. 1:28 calls for exertion of the spirit. Nature is to be the scene of our conquest and transformation. This understanding of American landscape according to the first Genesis account and the consequent moral mandate for work developed into the idea of progress.[31]

American nineteenth-century painting played just as major a role in the elaboration of this religious view as it did for its opposite. The dawn of the new age, discussed by Eliade and others, was definitively portrayed by Joshua Shaw in his *Coming of the White Man* (c. 1850; see fig. 3.13). The scenario that unfolds here is clearly the interpretation of the European-American Christian as Adam. The new Adam arrives by way of a glorious light, that is, with divine sanction, before which the Native Americans fall back, awestruck, over-

Figure 3.13. Joshua Shaw, *Coming of the White Man, c. 1850.*
Oil on canvas, 25" x 36". Carl Shaefer Dentzel, Northridge, California.

come by the light and power. Here we find the counterpart to the
expulsion from paradise: the course is now reversed in the new
arrival at the second opportunity for earthly paradise.

Since it faces us, requiring our moral effort and material change,
the wilderness provides both the clarifying challenge that defines
our task and the means whereby that task can be accomplished.
Thus, the natural doubly beckons and gives way. Emanuel Leutze
represents the scenario for the continent in his *Westward the Course
of Empire Takes Its Way* (1861; see fig. 3.14). The work was a national
icon of progress painted on the eve of the Civil War in the capitol
at Washington, D.C. We find our prospect melodramatically displayed as a sun-drenched promised land. Henry T. Tuckerman, the
"American Vasari," described the painting in 1867:

> An emigrant party, travel-stained and weary, who for
> long weeks have toiled in the face of interminable
> difficulties over the vast plains on the hither side of the
> Rocky Mountains, have reached, near sundown, the

Figure 3.14. Emanuel Leutze, *Westward the Course of Empire Takes Its Way, 1861.* Oil on canvas, 33¼" x 43⅜". National Museum of American Art, Smithsonian Institution, bequest of Sara Carr Upton.

> point whence the waters flow in the direction they them-
> selves are going, and from which they catch the first
> glimpse of the vast Pacific slope—their land of promise.
> El Dorado, indeed; for the earth and sky and mountain
> peaks are bathed in the golden glow of the setting sun.[32]

The meaning is further glossed in the typological construction of the central figures as the "Holy Family" and in the border, where we find Daniel Boone and Captain Clark framing a view of California's Golden Gate, the physical goal of the march.

Leutze paints a vision of manifest destiny like that which William

Gilpin ardently extolled in midnineteenth century in a rapture of nationalistic energy:

> The untransacted destiny of the American people is to subdue the continent—to rush over this vast field to the Pacific Ocean—to animate the many hundred millions of its people, and to cheer them upward . . . to establish a new order in human affairs . . . to regenerate superannuated nations—to change darkness into light . . . to teach old nations a new civilization—to confirm the destiny of the human race—to carry the career of mankind to its culminating point—to cause stagnant people to be reborn—to perfect science . . . to unite the world in a social family—to absolve the curse that weighs down humanity, and to shed blessings round the world! Divine task! Immortal mission! Let us tread fast and joyfully the open trail before us! Let every American heart open wide for patriotism to grow undiminished, and confide with religious faith in the sublime and prodigious destiny of his well-loved country.[33]

It was in America, as Gilpin put it, that the preeminently divine gifts "had been vouchsafed to the *American people* by God *through nature.*"[34] He said, "I discern . . . a new power, the *people occupied in the wilderness,* engaged at once in extracting from its recesses the omnipotent element of *gold coin,* and disbursing it immediately for the *industrial* conquest of the world."[35]

American Progress, or *Manifest Destiny* (1872; see fig. 3.15), by John Gast portrays just this preoccupation. The figure of progress presides over and guides the phases of our movement from east to west, not incidentally carrying a schoolbook in her right hand and a telegraph wire in her left. Before her storm clouds retreat toward the Rocky Mountains, while below her a stagecoach, pony express rider, transcontinental railroads, and even New York's Brooklyn Bridge are glimpsed. All the phases of subordinating the earth are represented.

The sublimity of nature lies, in this interpretation, not in the wilderness itself but in the scene as promise. Converting wilderness into paradise requires physically subjugating nature. For that task

Figure 3.15. John Gast, *American Progress*, or *Manifest Destiny, 1872*. Oil on board, 12⅛" x 16⅛". Gene Autry Western Heritage Museum, Los Angeles.

machinery is needed, especially machinery for communication and transportation, and hence we find the divinely sublime augmented with the machine.[36] The technological sublime emerges, as described in 1829: "The rudest inhabitant of our forest . . . is struck with the sublime power and self-moving majesty of a steamboat;— [he] lingers on shore where it passes—and follows its rapid, and almost magic course with silent admiration. The steam-engine in five years has enabled us to anticipate a state of things, which, in the ordinary course of events, it would have required a century to have produced."[37]

Artists enthusiastically portrayed the technological conquest of wilderness during midcentury. For example, in Thomas Rossiter's *Opening of the Wilderness* (c. 1846–1850) a newly constructed railroad depot is inserted into the virgin wilderness with, in the middle of the scene, powerful railway engines fired up, lights aglow, smoke streaking into the sky (see fig. 3.16). The same meaning is explicit in

Figure 3.16. Thomas Rossiter, *Opening of the Wilderness, c.1846–1850*. Oil on canvas, 17¾" x 32½". Bequest of Martha C. Karolik for the M. & M. Karolik Collection of American Paintings, 1815–1868. Courtesy, Museum of Fine Arts, Boston.

the title of an 1842 book, *The Paradise within Reach of All Men, by Power of Nature and Machinery.*[38]

Jasper Cropsey's popular *American Harvesting* (1864; see fig. 3.17) presents the story of the movement of the nation's chosen people: in the foreground are the stumps resulting from the work of the recent past; in the midground we find the beauty of success, accomplished at present; beyond, in the distance, the land yet to be cultivated, which awaits as future. The comparison with Cole's *Home in the Woods* is telling. Cropsey, undoubtedly aware of Cole's painting, attempted to surpass him, carrying the bliss of Cole's farmer to consummation. But the deeper story is found in the subtle difference between the works. Whereas Cole showed little evidence of agricultural labor or the passage of time, with his pioneer family living directly from the bounty of the natural garden, in Cropsey's version the passage of time through the three spatial and temporal zones (past, present, and future labor) provides the framework for the spacious barn and house, the village with church steeple, and ships of commerce. Although the central image is one of contained fertile paradise, that paradise obviously has been carved out of nature by human effort, for example, by the harvest of grain portrayed in the

Figure 3.17. Jasper Cropsey, *American Harvesting,* replica painted by the artist, 1864 (original, 1851). Oil on canvas, 35½" x 52¾". Indiana University Art Museum, Bloomington. Gift of Mrs. Nicholas H. Noyes.

center of the painting.

Although soon to be secularized almost beyond recognition in the materialist, economic interpretation of progress, the religious motivation underlying the progress depicted here is not yet obscured. Plainly, *American Harvesting* remains in the tradition of depicting our mastery of the wild and, although subtler than earlier works, does not depart from the common dream found in the folk tradition, for example, as embodied in Edward Hicks's *Residence of David Twining in 1785* (c. 1845–1846; see fig. 3.18).

Hicks's painting, in fact, portrays the farm where some sixty years earlier he had lived as an adopted orphan. While Mr. Hicks presides over the farm and its activity, Mrs. Hicks reads the Bible with young Edward. Success through control of the environment is a constant of this vision of mastery: Hicks shows the success of the mathematical-technological culture even in this "naïve" portrait of the rural

Figure 3.18. Edward Hicks, *The Residence of David Twining in 1785,* *c. 1845–1846.* Oil on canvas, 26" x 29½". The Carnegie Museum of Art, Pittsburgh. Howard N. Eavenson Memorial Fund, for the Howard N. Eavenson Americana Collection.

ideal and the results of industry. Here we find the achievement of prosperity promised in the Bible, wrought through action and depicted in the parallel bands of fence and furrow that compose the space. Again, space is made into home.

Indeed, so secure was the idea of progress through work that the natural world was often depicted as complicit. In the *Moll Map of North America* (1715) an amazing detail of the view of Niagara Falls shows "the Industry of the Beaver of Canada in making Dams to stop the Course of a Rivulet, in order to form a great Lake, about which they built their Habitations," wherein beaver work in orderly lines that would inspire any European rationalist.[39]

In extending the idea nearly a century later, Cropsey updates the

Figure 3.19. Jasper Cropsey, *Starrucca Viaduct, Pennsylvania, 1865.*
Oil on canvas, 22⅛" x 36⅜". The Toledo Museum of Art. Purchased
with funds from the Florence Scott Libbey Bequest in memory of her
father, Maurice A. Scott.

success of domestication by technology. For example, in *Starrucca
Viaduct, Pennsylvania* (1865; see fig. 3.19), he celebrates the prosperity achieved in the Susquehanna Valley in northeastern
Pennsylvania. Indeed, he portrays what appears to be a harmony in
wilderness transcended or cultivated, that is, in what has become a
landscape. From the pulpitlike rock in the foreground the spectators
admire the scene of a densely settled agrarian landscape showing in
great detail men working in harmony in a beneficent nature.
Farmers plough their fields, workers repair an old wooden bridge,
and smoke rises from the prosperous village. This desired transformation of wild nature into cultivated paradise is depicted in the rich
autumnal colors of fulfillment and harvest and is shown to be
accomplished by the material progress of train bridge and village; it
exemplifies Cropsey's quotation, some time earlier, of the psalmist's
statement that "the heavens declare the glory of God and the firma-
ment his handiwork" (Ps. 19:1).[40]

The religious interpretation of American nature as a paradise to
be wrestled by work from the wilderness came to an end as abrupt-

ly as its alternative. In this case, however, it was not that the idea of paradise was forgotten because a vision of the Garden of Eden fell into oblivion before the march of secular, material science. The interpretation of American nature as the promised paradise to be achieved through the transformation of wilderness fell into obscurity in the course of the myth's success. That is, the success of technological progress itself displaced its own foundation: religious understanding gave way to material accomplishment.

American nature was no longer a scene for a religiously understood mission because the radically secular view, which at first depended on and was fed by a religious understanding, gained enough power and became so taken for granted as the assumption of regular activity that it surpassed its source and thus overlooked it. Regardless of the details of the eclipse, by the time the frontier had fallen into settlement, the religious understanding had played out, absorbed into and then dissipated into secular beliefs in progress. What had been a landscape of religious mission disappeared into the secular landscape, so physically understood that the frontier itself became thought of as the kind of phenomenon that could officially be declared closed by the superintendent of the census—an amazing transformation of meaning, to which we do not give a second thought because it now seems so obvious to us.

Secular Echoes in Landscape Architecture and Environmental Attitudes

Precisely because the secular economic and political understanding of America as destined for settlement and endless progress is the more familiar and even taken-for-granted account, we need to keep in mind the original religious motive. What by now may appear to be an uncontroversial justification for material progress conceals its origin and power by that appearance. That is, the religious interpretation of nature, substantially accomplished by artists, becomes so powerful, so widely shared, that it finally becomes invisible, so taken for granted that it ceases to be focally operative. In the end it dissolves in the achievement of its goal, its own success.

Against this dominating vision it is even more surprising that the alternative vision of Cole, Church, and Bierstadt should have flowered. That the American landscape generally was seen in religious terms seems odd to us now, so lost is the understanding and tradition. How more deeply lost is the subtler, inner struggle between two traditions, both derived from Genesis.

Nonetheless, the accomplishment and alternatives, even if forgotten, remain with us, perhaps more powerful than ever because they operate as hidden directives. Consider what we assume about disputes in the territory and practice of environmental interpretation: we worry about whether the approach of conservation or stewardship should prevail; we take it as obvious that notions of development as represented, say, by the views of conservationists and developers are wildly opposite. Actually, in these cases we face once again an old dilemma. Is wilderness or what now seems natural a last fragment of a garden of Eden, which we should let be, preserve, and try to enjoy as it is, or is it raw material for the work of civilization, which will domesticate and transform nature into a garden?

In the first case we are called on to change our dominant view and transform ourselves so as to be worthy to enter the natural paradise or garden given to us. We would try to overcome the attitude that our action should be based on self-confidence and personal desires, for that attitude earlier led to our expulsion from paradise and now would destroy our second chance. In the second case we are challenged to complete the mastery and transformation of the natural into the cultivated paradise, which is the goal of our work and requires the mature acceptance of our responsibility for our redemption in conjunction with what is given to us. In either case, although with our little knowing it, we are guided by the alternative interpretations already developed in the nineteenth century with considerable sophistication.

If we fail to take into account these origins of our unconsidered attitudes and approaches—long since radically secularized—we cannot, then, think or act as responsibly as we might. Indeed, our currently assumed attitudes remain confused as long as we are oblivious to both their real differences and the way in which they are unified as variations on a theme (as the two accounts in Genesis differ yet belong together in our complex religious tradition). Recovering and attending to this forgotten origin of our interpretation is crucial for critical self-understanding and for responsible action in the landscape.

In addition to helping to clarify our own current positions by disclosing hidden meanings and assumptions as outlined above, recovering the original interpretations would have other practical consequences. First, it could contribute, from an overlooked source, to the current questions about the origins of our Western environmental attitudes. For example, although there is considerable debate about the extent to which Christianity lies behind ecology,[41] little attention has been paid to the way in which art history contributes to the issue. Perhaps a better way to put it is that we have not drawn, all at once, what we know in theory and history of landscape, ecology, art history, and history of religion and culture into a focal and integrated environmental interpretation.

Second, recovering the complex religious interpretation of the American landscape in the nineteenth century—especially since that interpretation grounded the understanding not only of our American landscape but also of the meaning of American nature and the United States itself—calls for reconsidering the history of American landscape architecture. No one would deny that the development of an American landscape architecture tradition, for example, in our gardens and parks, is partially a consequence of European events and attitudes, such as English "romantic" gardens. Nevertheless, the struggle to develop a uniquely worthy—and inherently religious—view of our own land, which was a finally successful motive for nineteenth-century landscape painting, also involved a broader struggle to overcome European conventions, as well as a complex set of historical assumptions. The unique accomplishments, then, of American painters and the development of American perceptions beyond European conventions surely need to be taken in account in understanding our own development and unique landscape.

As a simple example, since England never was understood as a second paradise, whereas America was, an American park (the outcome of the tangled interpretation of landscape as garden of Eden and site of material transformation, all somehow secularized) cannot mean the same as an English park that might appear to be similar.[42] In short, if we take seriously that we have developed an indigenous understanding of American landscape, partially generated by nineteenth-century landscape painters, then that tradition should no

longer be overlooked by the theory and history of landscape architecture, which say little or nothing of the fundamental religious
understanding as a generative force.

That the contemporary landscape echoes our earlier religious
understanding is almost entirely obscured. Yet current attitudes
toward parks and gardens—indeed, the meaning and function of
urban open space today and the principles of design and planning—
are derived from that tradition. At the least, our understanding of
earlier landscape architects and the history and vocabulary of landscape architecture would need to be rethought and, likely, revised.

The connection (and remaining influence) may be seen in a
figure whose work and writings we continue to use in contemporary
theory and design, although in adapting Frederick Law Olmsted
(1822–1903), we largely are unaware of the complexity of what he
said and what his landscapes meant.

To cite one issue: where in the spectrum of this tradition (within
the poles of secular and material vs. religious interpretation and of
European vs. uniquely American conventions) was Olmsted operating in the development of the nineteenth-century urban landscape?
Even in the best scholarship, the question is too little asked or, when
noted, not really worked out.[43]

In fact, Olmsted is a figure at the turning point in our shifting
interpretations of the American landscape. He participated in all
three of the previously presented traditions: he understood nature as
an earthly paradise given by its creator; he saw nature as the site for
our moral industry that would be rewarded by the fruits of the
promised garden; and he passed out of any deeply religious understanding of nature to help to develop the scientific approach to environmental planning and human well-being.

Although it is something of a commonplace that our national
parks, especially Yellowstone and Niagara Falls, were established
partially through the influence of the paintings of artists such as
Church, Bierstadt, and Moran ("the father of parks"),[44] the relation
between the theme of natural divinity in their attitudes and work
and its appearance in Olmsted's largely is overlooked. Nonetheless,
Olmsted's efforts to preserve Yosemite and Niagara Falls in a state as
close as possible to the natural seem to have been grounded in the
interpretation of nature as a divinely given paradise contrasted with

the work of humans.

Indeed, it was Church himself who, in 1869, first called the destruction of the scenery of Niagara Falls to Olmsted's attention, influencing Olmsted to help establish a reservation there (architect Henry H. Richards was also one of the petitioners).[45] Since Church (as well as his contemporaries) had represented Niagara Falls as a sign of the biblical deluge and as a symbol to America of God's promise, Olmsted would have been aware of the connotations of the project.

As we have seen, a shared "prejudice" of the time was that what had been prefigured in the Old World and Old Testament was brought to fulfillment in the New World. So accepted was the convention that the visitors who flocked to Niagara Falls after the Erie Canal opened in 1826 not only saw the place in utilitarian terms or with the amateur's keen interest in natural history but frequently experienced the religious sublime. The pilgrimage to the falls not uncommonly led to rapture. For Harriet Beecher Stowe, the falls evoked images from the Book of Revelation. Another woman's letter testified, "The roar of the waters agitated me. . . . I cannot sooth down my heart—it is kindled by deep works of the invisible. . . . A great voice seems to be calling on me. . . . I have felt a spell on my soul as if Deity stood visible there . . . I felt the moral influence of the scene acting on my spiritual nature, and while lingering at the summit alone, offered a simple prayer."[46]

Church's *Niagara* (1857) participates in this horizon of meaning, immediately engaging viewers, who on looking at the painting, which has no foreground, find themselves as if suspended directly over the water, almost at the brink of the falls (see fig. 3.20). In Church's dramatic presentation the water powerfully and majestically falling over the full horseshoe (seen from the Canadian side), for all its immediacy, yet occurs under nonthreatening skies. Heavens and earth are united by a rainbow that dynamically arches from the sky in the upper left of the painting down into the precipice in the center. Surely here we have an image of a landscape where God confers his blessing on Americans.

Since this *Niagara,* one of his several treatments of the subject, "was the picture that made Church the most famous painter in America,"[47] Olmsted would certainly have known both the painting

Figure 3.20. Frederic Edwin Church, *Niagara, 1857.* Oil on canvas, 42½" x 90½". (107.95 cm x 229.87 cm). In the collection of the Corcoran Gallery of Art, Museum Purchase, Gallery Fund, Washington, D.C.

and its religious interpretation. In this context, when Olmsted advocated a plan "to preserve and develop a particular character of natural scenery on a great scale avoiding as much as possible all manifestation of art, human labor, or human purposes,"[48] the argument implicitly, but obviously, drew on the spiritual foundation of the understanding of the landscape and advocated that basis as a dimension of what became the national parks.[49]

Olmsted's awareness of the tradition underlying Church's work (or "natural paradise as given," as I have called it in this chapter) is apparent from Olmsted's personal history and writings. Indeed, Olmsted's fluency in the basic edenic idiom continues throughout his life. For instance, in 1866 he described the goal for the campus at Berkeley, approvingly quoting Lord Bacon,

> who three hundred years ago, sagaciously observed: "God Almighty first planted a garden, and, indeed, it is the purest of human pleasures; it is the greatest of refreshment to the spirits of man, without which buildings and palaces are but gross handiworks: and a man shall ever see that when ages grow to civility and elegance, men come to build stately sooner than to garden finely—as if gardening were the greater perfection."[50]

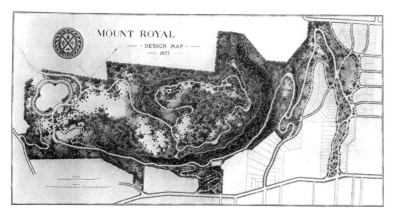

Figure 3.21. Frederick Law Olmsted, *Design Map for Mount Royal, Montreal, 1877*. Courtesy: National Park Service, Frederick Law Olmsted National Historic Site.

Later in this report Olmsted elaborated the point that such a garden is the scene of the promised paradise and thus the site for our mandated moral work: "[It is] certain that if [townspeople] fail to secure fresh air in abundance, pleasant natural scenery, trees, flowers, birds, and, in short, all the essential advantages of a rural residence, they will possess but a meager share of the reward which Providence offers in this world to the exercise of prudence, economy, and wise forecast."[51]

Similarly, he advocated in clearly religious rhetoric drawn from the second Genesis account that "wild gardening" should be protected in Mount Royal Park in Montreal (see fig. 3.21):

> I will go further, and tell you that if you cannot afford to keep a single man so employed [as a gardener], there are hundreds of little places on the mountain within which, if you can but persuade yourselves to regard them as sacred places and save them from sacrilegious hands and feet, the original Garden of Eden will delight your eyes with little pictures within greater pictures of indescribable loveliness. And remember that it is the *lilies of the field, not the lilies of the garden* we are bid consider.[52]

That Olmsted should write this way as late as 1881 leaves little doubt that it was a mature consideration.

Still, such overt rhetoric seems to have been exceptional. By and large Olmsted operated within the more dominant tradition where American land was understood as the scene for our work in transforming wilderness into a moral world (following the first account in Genesis). Even more important, and showing the rhetorical client- and user-oriented functions such religious language finally served in the last part of the nineteenth century, Olmsted, like many of his urban contemporaries, rejected fundamentalism and organized religion for rational, secularized doctrines and their democratic social ethic.

Olmsted's background and circle of acquaintances further clarify how he and other early landscape architects were aware of and connected to the reigning religious understanding, even as they moved away from it. Olmsted was intensely exposed, although apparently in rather forced circumstances, to orthodox religious instruction from age six to twelve, while he was in the care of six successive ministers at one fundamentalist boarding school after another.[53]

As a young man he participated in a "period of religious seriousness among [his] friends in New Haven during the early months of 1846."[54] At the time he not only taught Sunday School but thought through his religious position in discussions and in a series of letters with friends, especially Charles Loring Brace, who introduced him to the doctrines of Congregational minister Dr. Horace Bushnell. Olmsted's and Brace's friendship developed initially between 1842 and 1845, when Brace shared rooms with Olmsted's brother, John Hull, at Yale. Olmsted later accompanied his brother and Brace on their 1850 walking tour in England.[55]

Writing to Brace in 1846 (and mentioning the daguerreotypes of themselves that the "five friends" had made; see fig. 3.22), Olmsted discusses the theology of Bushnell and William Ellery Channing, on one occasion noting that he will complete a letter in progress "whenever I happen to feel in a Metaphysico Theologo humour."[56] In fact, he discussed religion, social reform, and aesthetic ideas with Bushnell himself on at least one occasion (Bushnell lived next to Olmsted's parents in Hartford from 1836 to 1841, and Olmsted's father joined his church in 1848).

Figure 3.22. The five friends in New Haven days, 1846. Back row, left to right: Charles Trask, Frederick Kingsbury, John Hull Olmsted; front row: Charles Loring Brace, Frederick Law Olmsted. Courtesy: National Park Service, Frederick Law Olmsted National Historic Site.

With Bushnell we have a specific case that confirms the explicit union of nineteenth-century attitudes to the landscape and city planning with self-conscious biblical interpretation. The Reverend Bushnell instantiates the unity: author of a major nineteenth-century essay on city planning, "City Plans," he was also, as author of works such as *Nature and the Supernatural, as Together Constituting the One System of God,* one of the most influential Protestant theologians and biblical interpreters of his time.[57] His interpretation according to the Bible and Christian symbolism remained within the tradition of hermeneutics I elaborated in the introduction; at the same time it was part of the midnineteenth-century transition from religious to secular and aesthetic views and social reforms.

In developing his radical theories of religious language and symbolism, Bushnell explicitly took up the issue of nature, deity, and Scripture. Although he held that the world is a source of inspiration bringing us into the presence of divinity, he also argued that transcendent reality is not transmitted only by nature or just anywhere; rather, nature is seen through culturally developed religious language and symbolism.[58] That is, Bushnell stressed the "perceptive and aesthetic dimensions of faith."[59]

As to social reform and the environment, Bushnell argued for restoring nature to the cities, actually laid out a park for Hartford, and called for professional urban planners. In "City Plans" he wrote that the welfare of a city and its inhabitants "depends, to a considerable degree, on the right arrangement and due multiplication of vacant spaces" and "the providing and right location of a sufficient park, or parks," which would provide "breathing places."[60] It is little wonder that Brace, when he founded the Children's Aid Society in New York, placed slum orphans in rural homes and Olmsted, moving in the complementary direction, worked to bring nature into the city, although the religious motive and force of the connection would soon be forgotten by others.

To refocus all this: after growing up amid these religious figures and after maturely considering the ideas, Olmsted certainly understood and was fluent in the dominant theological interpretations of nature. Because he could not reconcile his religious views with required practices and belief systems, however, after the intense year of 1846 he chose not to make a profession of faith or join a church.[61]

Nor was Olmsted unusual in this regard. His friend Brace, who had gone on to become a divinity student, also "found that his purely theological interests were waning and hoped that his travels [in 1850] would help him prepare for a career of Christian service in social work."[62]

Thus, as an adult Olmsted belonged to a group of liberal Protestant leaders and their influential disciples who were in the forefront of the movement from formal religion to secular doctrine and social responsibility, especially urban reform. Clearly identified with established social, economic, and political power in the New York area, they operated with an idealism based on Bushnell's organicist thought and on "radical Protestant theology as expounded by the Unitarian minister William Ellery Channing and by his followers, [minister Henry Whitney] Bellows and [newspaper editor William Cullen] Bryan."[63]

Of course, Olmsted, Brace, and their colleagues operated at the end of the tradition I have examined, when divine providence was interpreted as progress. That is, at heart, for Olmsted's circle, the landscape no longer recalls God through his creation as it did in the primary sacred tradition: even religiously based civic ethics and humanitarian goals are at a far remove from sacred disclosure.[64] Now American nature is seen as a means to human well-being in the context of manifest destiny. As Albert Fein observes, the New England church and common were replaced not so much by urban Gothic cathedrals as by parks.[65]

Olmsted actively and creatively participated in the development of landscape and parks as a part of the movement to create a homogeneous and harmonious national urban society. As one of the group, Parke Godwin, held, "Providence" had specified "this continent, and the people, for a homogeneous civilization."[66] Olmsted displays the same religiously couched assumptions and language in explaining the benefits of parks:

> Consider that the New York Park and the Brooklyn Park are the only places in those associated cities where, in this eighteenth hundred and seventieth year after Christ, you will find a body of Christians coming together, and with an evident glee in the prospect of coming together,

all classes largely represented, with a common purpose, not at all intellectual, competitive with none, disposing to jealousy and spiritual or intellectual pride toward none, each individual adding by his mere presence to the pleasure of all others, all helping to the greater happiness of each.[67]

Central Park, for example, according to Olmsted, "exercise[s] a distinctly harmonizing and refining influence upon [even] the most unfortunate and most lawless classes of the city."[68] In his view urban parks and gardens operate as sites for humanitarian improvements and the development of a unified, democratic civic realm.[69]

At this point, despite supporting the value of natural scenery in national parks, Olmsted and others, by and large, no longer envisioned the natural as wilderness but viewed it as the obviously cultivated. In one way this natural, "rural" open space was the same as the domesticated spaces that landscape painters such as Cropsey delineated as subjugated and cultivated.[70]

At the same time, Olmsted had come to hold that the countryside, with its abysmal conditions, was a failure and that the future lay in the urban environment.[71] This change in attitude was part of a cultural shift in which the idea of an agrarian society was giving way to the rise of the cities. Olmsted and his colleagues sought to fuse the best characteristics of both realms. In their union these two opposites would be transformed into complements: the derelict and disadvantaged countryside was to be replaced by the cities with their amenities; the cities would be reformed and made well by planning and open space. The antiurban force would be assimilated into the city, so that the city, brought to its full health and potential, could "witness to man's spiritual destiny," as William Ellery Channing put it.[72]

The strategy, then, was for nature (cultivated "rural" open space) to be brought into towns, so that each's virtues would cure the other's vices. For example, the benefits of the natural, such as fresh air and a mixture of rest and exercise, provide the antidote to the city's ills. That is why Olmsted specified that parks were to be built to hide the city and, as its counterbalance, to soothe us with their healthy environment.

For all these religious dimensions, however, even in their civic manifestations, Olmsted does not remain within a deeply religious interpretation of the landscape, because he passes over to advocate and develop the secular scientific approach to the environment and progress in confluence with social Darwinism—although not without differences and not as materialistically as Clarence King.[73] Indeed, this is how we best know him, in the forefront of the scientific approach that was displacing the old religious visions. His stance clearly was not a simple matter of personal indifference to religious orthodoxy but the result of his convictions and of his contributions to the science of environmental planning. He was not concerned with tradition's religious goal of paradise either on earth or in heaven; rather, he strove toward humanitarian reform through scientific planning and design of the urban environment (and to a much lesser extent, national parkland).

Parks and other urban open spaces, for Olmsted, were to be designed to promote social well-being: health, decency, vigor, civil moral tone, sensibility to the beautiful, trade, and prosperity—all humane goals of environmental and social planning.

In short, Olmsted represents landscape architecture and planning at a transition point. He participated in both traditions of the religious mind. With Frederic Edwin Church he was able to advocate letting nature be, "naturally" (for example, in national parks), without human improvement. With the heirs of dominant Protestantism, such as Horace Bushnell—who, despite increasingly developing liberal moral and civic progress and its humane rewards, still believed in religious responsibility to work to improve the world morally—his planning and the design of urban parks were intended to improve the city and landscape. Ultimately he moved beyond these two strands of the old religious traditions and on to the secular scientific approach to the environment and the planning and design of urban open space as we know it today.

Because of this complexity and transition, Olmsted himself needs to be better understood. He was not ultimately an advocate of the older sacred understanding of the landscape. He was a very skilled politician and environmental planner. At times his use of religious language and symbolism seems to result from calculated professional strategy. Olmsted the masterfully deft politician surely referred to the Garden of Eden and Genesis in the plan for the Berkeley campus because the chairman of the committee at Berkeley was the *Reverend* S. H. Willey. Nonetheless, fully understanding Olmsted requires at least understanding how the diverse strands of thought and language functioned in his life and work. The previously given sketch of his participation in the several dimensions of this tradition and in the transition from it is only a first step toward recovering what his work means and what we can learn from his continued use of religious rhetoric late in life.

To step back to our larger concern with environmental interpretation—since the original sacred interpretations of landscape passed over, hidden, into secular, civic planning and environments—the issue does not really focus on Olmsted. He is only an interesting and pivotal figure. In addition to having individual importance and influence, Olmsted is symbolically significant: he represents and presents to us today a substantial set of issues yet to be resolved in our environmental interpretation. His work indicates how the lost sacred understanding of landscape is important for understanding current scholarly and design issues. No matter what pragmatic context is noted in his writing, the rhetoric and fundamental environmental principles unmistakably echo an edenic social-religious harmony. Hence, the issue remains of whether the development of American parks, for example, can finally be understood without thinking through their place in the secularized religious interpretation of American landscape—especially if we wish to recover and preserve what appears to be a unique landscape tradition.

The hermeneutic of the landscape complements traditional histories of the landscape by partially disclosing the "hidden" dynamic behind the emergence of the American environment. The story of our existing parks, even when told in so sophisticated and astute a work as Galen Cranz's *Politics of Park Design*,[74] is incomplete and depends for its deeper interpretation on the landscape's concealed origin in religious attitudes and their subsequent transformation into a secular political and civic view.

Cranz's analysis convincingly begins with the idea of the park as a "pleasure ground" and goes on to account for the contrary "reform park" and the later forms of "recreational facility" and "the open-space system" (see fig. 3.23). The pleasure ground, which offers a respite from the city because of the soothing and restorative power

Figure 3.23. Nineteenth-century park as pleasure ground. Boating on the lagoon, Lincoln Park, Chicago, 1890. Chicago Historical Society (ICHi-03420).

of nature, gave way to the goals of forming civic character and unity as part of the zealous moralistic reform and progressive attitudes that came to dominate by the early 1900s. The natural, which Olmsted and others once understood as the opposite of the city and antidote to its ills, by the turn of the century became an instrument of the city. The park and garden became the spaces for works of moral vigor and education especially as heterogeneous and non-English-speaking immigrants were ordered, kept out of trouble, and "Americanized" (see fig. 3.24). Hence, as Cranz argues, parks became an instrument of discipline designed to organize and routinize social attitudes and beliefs.

These basic forms and their historical dynamic need to be questioned within the horizon sketched out here. Today's efforts to understand and design appropriate open spaces have much to learn from the nineteenth century if we critically inquire into (1) how the unstructured enjoyment of the pleasure ground (itself entirely con-

Figure 3.24. Twentieth-century reform park. Settlement kindergarten class at Davis Square Park, Chicago. Chicago Historical Society (ICHi-03380).

structed and not at all "naturally given," though it may so appear) derives from the prior interpretation of the garden as a paradise already given for Americans to enter and delight in and (2) how the reform park (appearing as more built than natural) results from interpreting nature as the site for the transformation of our land and our lives through moral effort. (Or perhaps, although it is a less likely account, through Olmsted and his colleagues, the dominant American ideology already had eliminated the first tradition and had selected the second so that nature was seen as the cultivated and the scene of formation of civic character, regardless of whether in a more relaxed or more disciplined manner.) In either case, from Olmsted on, American parks have retained their origin in nature understood through paradise, since American designers inevitably implemented, opposed, and unconsciously adapted his principles and since his circle's beliefs and work—perhaps unintentionally—led to the moral reform and progressive attitudes that dominated when the reform park replaced the pleasure ground.

For those interested more in design than in scholarship, the issue remains important. As Galen Cranz observes, "I soon learned that I needed to understand the ultimate purpose of parks in order to design . . . playgrounds."[75] If we do not understand how and why our parks and open spaces are what they are, if we do not understand the meanings of the elements, or wholes, or the place of landscape in the urban context, design will adapt the past unthoughtfully, uncritically, and eclectically. That is, new design will operate without adequately understanding its own uniquely American tradition, vocabulary, and possibilities—and continue to be an unwitting instrument in our own cultural displacement.

The Hidden and Disclosure

Nature as paradise is understood here entirely in terms of Christian eschatology (although other religions such as Judaism and Islam share some of the same concepts). We might say that this was known all along, even if we forgot it for a while. That is, it was known at the time and now may be recovered, which is all to the good for historical reasons. But is that adequate to fulfill the claim that hermeneutics recovers deep meanings truly hidden? What was so hidden if everyone once knew it and we now know it again?

In the first place, calling a relation to our attention through patient and detailed historical research does not mean that we now understand it. The meaning of the twinned Old Testament roots, as interpreted by Catholic and Protestant theology, is profound if we begin to think it through, a task that goes far beyond merely noticing, and that will take some time. Within this recovery we find that the twinned attitudes from Genesis vie with each other without being able to resolve themselves into a settled view. This internal dynamic needs to inform reflections on the religious origins of American attitudes to nature and, today, to ecological perspectives.[76] A revisionary hermeneutics has only begun to recover our uniquely American landscape tradition.

Second, this theological interpretation has guided all other interpretations of nature, translating them into its own terms, overtaking and consuming them. Ubiquitously known and used during the nineteenth century, its power obscured all other alternatives. It can be objected that the other alternatives, such as the views of native peoples, were acknowledged, not ignored, but that was not so in any significant sense. Indigenous peoples' truths were denied from the start as they were assimilated through the Christian master terms. The ontotheological basis of both Christian traditions derived from Genesis interprets God as creator and the natural world as his creation.[77] This is not a general idea but a specific way of interpreting reality derived from Platonic and Aristotelian metaphysics. Nature is the set of beings that derives its existence from the original and continued outpouring of Being, where Being is understood as God or as a primary manifestation of God according to natural theology.[78]

Thus, acknowledging native peoples' views of nature amounted to taking their "parallel" accounts as "creation" myths, myths that were judged and understood in the light of the master creation story in Genesis and of the concepts and categories of two millennia of theology.[79] The native accounts could be understood as childish, defective, or ignorant because they did not really see who God, as the creator or as ultimate being, was or how he brought about and sustained creation. Whatever in these native accounts did not fit into the master terms of creator and creatures could be dismissed or attributed to stubbornness in not accepting the true version.[80]

As a variation, the Native Americans' attitudes to nature as some sort of living whole were translated straightforwardly as pantheism, a "well-understood" ontotheological phenomenon. Judged metaphysically this indigenous version misunderstood or confused the difference between beings and being and so was philosophically incorrect. Theologically it erred in confusing the created with creator and accordingly was heresy. Thus, the apparently obvious revelation of the American landscape as God's graceful blessing obliterated all other traditions of interpretation and possible ways of life. The latter could not possibly have conveyed their own truths because they were not allowed to speak in their own voices.

The indigenous traditions referred to, however, were not at all homogeneous or unified (in 1492 there were more than four hundred languages and ways of life in what became the United States and Canada), yet our very way of referring to native peoples assumes that somehow they were fundamentally the same. Their apparent identity was constituted by default since each group was different from and deficient in regard to the master principles established by the Christian theological account and the conquerors' identity. Because the alternative accounts that specific tribes and languages offered were judged to be inconsequential compared to "the truth," all secondary differences among the former were erased and deemed inconsequential.

Here we also glimpse a third level of what was hidden by the nineteenth-century understanding. In this tradition, as we have seen, nature appears as landscape, as something viewed. Although how nature came to be seen as a landscape by Western culture is a long story in itself, the essential features of this process involve the ontological and epistemological opposition of subject and object.[81] Insofar as a subject is understood as consciousness, standing over and against the natural object, the natural can manifest itself as something that comes into view. Nature as viewed during a sight-seeing tour, through a Claude glass, or in poetic or artistic repre-sentations is a landscape. The entire aesthetics of nature develops its categories to explain the different ways that landscape shows itself, as sublime, picturesque, beautiful, and so on.

This view of nature as landscape is also at base a metaphysical, representational view. It initially is possible because God is under-stood as omniscient, as the all-seeing being who views creation and then evaluates it. During and after the Creation God's complemen-tary act was to stand back to size up the propriety and harmony of the world in terms of what he had willed: "And God saw everything that He had made, and behold, it was very good."[82] Since God is the creator par excellence, and since we are made in his image and like-ness, we exhibit our own natures, our own participation in divinity, insofar as we too are creative. Thus the soul and inspiration have long been part of the Neoplatonic interpretation of artistic creativi-ty, a tradition later modified in the modern secular era.

In all these theories nature is nothing in itself. It is something only insofar as it is the created. To the religious understanding it is significant only insofar as it reflects its divine origin or as, repre-sented and operated on in human imitation of divine act, it is trans-formed and completed by the hand of the artist or our pragmatic technologies. Here nature is always a sign or counter in the larger

play of divine creation and human re-creation, of divine presentation to us and our re-presentation to ourselves and to God, who is
always watching and receiving our prayers and offerings. Nature
really appears as a mirror, reflecting the acts and interactions of God
and his human companions. As the mirroring symbol in the middle, nature shows each to the other.

Nature symbolizes God and his grace to his chosen people.
Hence nature, which appears as the rainbow and as Niagara Falls,
must be interpreted through the Old Testament. Alternatively,
nature symbolizes our increased self-awareness and acceptance of
ourselves as blessed by God and as needing to act as he wills, a symbolism delineated and shared by the artist so that we all can come
to self-understanding and appropriate national action. Nature presented to us as landscape shows—represents—God to us. Nature
represented in our paintings, poems, and environmental works symbolizes to ourselves, and back to God, our comprehension and willingness to accept our essential relation and mission.

Nonetheless, all versions of nature as landscape—whatever the
differences among Plato, Plotinus, medieval theology, Renaissance
Neoplatonism, Catholic missionary views in the New World gov-
erned by France and Spain, Puritan attitudes to wilderness in New
England, nineteenth-century pietism, and the secularized scientific
heritage—are part of the same ontotheological (that is, metaphysical and representational) system. Nature as symbol is a kind of coin
that can pay various debts or play in various games. No matter what
the various transformations are, however, it has its meaning only in
terms of the master system and is "paid off" only in terms of what
the system ultimately values (blessing, salvation, participation with
God, truth, etc.).

Because of the appearance of nature as landscape in the Western
theological tradition, an even deeper obliteration occurs than that in
which the dominant creation story hides alternative indigenous
views. Here, nature as created necessarily appears as landscape, that
is, as the material medium between God and humans, where the
spiritual nature of the latter two is all that matters. Nature is
insignificant as material (compared with divinity and our soul, that
special spark of the divine that is treasured because it defines our
essence). Material nature is valuable only as the symbolic counter

that allows transactions between God and his chosen American people. Thus, in this cultural context, the possibility of nature as something in itself is fully obscured. Since nature is given as a symbolic object in the midst of a representational ontological-epistemological chain, any inherent meaning is out of the question. It never could be disclosed, never could come into appearance.

Of course, from our contemporary vantage point we can see that nature, as physical, is something in itself, at least to an extent, and more so for materialism than for humanism. That dimension manifests itself, however, only when science wrests it from previous religious interpretation. Nature can be seen as fully material only when it is divested of its spiritual horizon. The struggle we saw at the end of the story of natural divinity is nothing less than the conquest of the sacred by profane science, the process wherein nature disappears as symbol, that is, as the place of the interaction of God and soul. When this old view passes away, it is supposed that nature can be seen in itself. In appearing as a purely material object to scientific, technological human subjects, however, what we call reality still remains metaphysical.

The spiritual dimension has been removed, but the fundamental metaphysical structure remains. Subjects are no longer religiously understood, but secularized will, consciousness, and judgment remain the essential traits of humans. Objects are no longer understood as creation and sacred symbols but are secularized as the merely physical raw material for human projects. Consequently, nature is still defined in terms of its potential for being operated on by its masters and is valued only in terms of our aesthetic, social, and economic desires.

It is difficult even to think of what an alternative might be. What would nature possibly mean, understood as something in itself, if not the object of our positive knowledge and technology? An example, irrelevant here except that it is nonmetaphysical, is the early Greek *physis* that later became *natura* in Latin and eventually what we call "nature."[83] *Physis* is the self-generating and ever-coming, the energy of coming forth out of hiddenness and abiding, which goes on in bewildering, incredibly powerful ways that overwhelm and challenge humans to try to gain some localized mastery. Mastery, however, is never possible, because considered on a macrolevel scale we are an altogether insignificant part of *physis*'s upheavals, and on a microlevel scale, as mortal, we come forth only for the briefest moment. Even in this inadequately short sketch it becomes clear that the word *nature* is unnecessary here, an inappropriate translation after the fact that misses a large part of the meaning said with *physis*.

To return to my American example: nature might have been talked and thought about in a way other than as landscape or as a parallel metaphysical concept. For example, Luther Standing Bear, explaining the difference between the newcomers and his people, said of his world, "Earth was bountiful and we were surrounded with the blessing of the Great Mystery," an image depicted in his water-color *Black Elk at the Center of the Earth* (1947–1948; see fig. 3.25).[84] Note how hard it is, even here, to resist mentally translating this saying into the ontotheological tradition, that is, to free ourselves from immediately and unself-consciously reading and understanding "Great Mystery" as "God" through Judeo-Christian theology. English words fail here. Black Elk said, "Wingeds, the two-leggeds, and the four-leggeds, are really the gift of *Wakan-Tanka*. They are *wakan* and should be treated as such."[85] Although Luther Standing Bear and Black Elk seem to be saying something roughly congruent and that appears pregnant with implications for understanding and acting in the world, we certainly cannot presume that the two say the same thing. Therefore, it is an unjustified leap to think "blessing of the Great Mystery" alongside "gift of *Wakan-Tanka*."

Those learned in indigenous languages tell us that *wakan* roughly means "the sacred power that permeates all natural forms and movements."[86] There appears to be a cluster or family of words that say what we can provisionally think of as a power that vivifies things, where the individual things have little existence except while the power moves in and through them. Archaic Lakotah says *"śkan, taku śkanśkan*: something-in-movement, spiritual vitality."[87] Navaho says *"ałíʾ*; power; special, extraordinary power."[88] Zuni says *"miłi*"; Iroquois, *"orenda"*; Algonquian, *"manito"*; Shoshone, *"pokunt"*; Hidatsa, *"xupa"*; Athabascan, *"coen"*; Crow, *"maxpe"*; Pawnee, *"Tirawa"* (a more personalized form, as is the Dakotah/Lakotah *Wakan-Tanka*).[89]

Suppose we can hear the subtlest hint of how these words say that

BLACK ELK AT THE CENTER OF THE EARTH

Figure 3.25. Surrounded with the blessing of the Great Mystery. Luther Standing Bear, *Black Elk at the Center of the Earth,* 1947–1948. Watercolor on paper. Negative no. 337964. Courtesy, Department of Library Services, American Museum of Natural History.

what we understand as beings, objects, and events bear meaning and value insofar as they are given by and disclose a primal force. How could we begin to interpret the way in which any one of these groups experienced the force that comes and goes across the world and in which humans participate for a short time? What would the mystery that gives and grants, the mystery that is beheld on the earth and in the heavens, have disclosed within the worlds in and through which it occurred before it withdrew?

It would be absurd to pretend that any but a handful of scholars might understand such sayings or possibilities. We neither speak these indigenous languages nor participate in the living cultures where such words name the mode of the coming to be and passing away of local worlds. Nevertheless, we can at least ponder how *śkanśkan, wakan,* and hundreds of other words and phenomena as interpreted by indigenous peoples would have been part of the disclosure of "nature" as inherently and nonmetaphysically meaningful. Any attempt to translate such given words and occurrences into representational terms destroys the very subject matter. Such a move would collapse the project of interpretation in an instant, as happened with the anthropological conceptualization of *mana*.[90] Here we can only note the remote possibility that a hermeneutics might recover indigenous peoples' realms in their own terms, although that seems very difficult and unlikely. Of course, such an insight, if retrievable, not only would be of the greatest historical importance but would be a kind of earthquake into the future, shaking up and informing at least our meager attempts at an appropriate ecology, if not a once again spiritual attitude toward the earth, heavens, and all life.[91]

Apparently the alternatives to the religious, metaphysical interpretations of the nineteenth century were nominally acknowledged, although actually destroyed in translation and assimilation into the dominating system and thus finally hidden. The revelation of natural divinity was so fully a matter of metaphysical concepts (such as "language," "symbol," "spiritual subjects," and "material objects") that the complementary nonmetaphysical disclosures of mystery or power that might have appeared as inherently valuable were simultaneously obscured—as the sun necessarily obscures its shadow from itself. A final concealment discloses itself here.[92]

The nineteenth-century intellectuals, scientists, and artists who worked with tenacious success to disclose the world as they did, by articulating natural divinity and thereby shaping cultural understanding and practices, could not possibly think that theirs was but one passing manifestation of earth, sky, divinity, and humans, as Heidegger shows it to be. Nor does this observation claim any superior insight from the vantage point of anthropological relativism, which holds that different groups naturally see things, even nature, in different ways. That would be to reduce the accomplishment of the world's appearing as natural divinity to merely one of many arbitrary, more or less interesting perspectives. That also would relegate the possible, subsequently obliterated alternatives, such as those of indigenous peoples, to appearing as so many points of view. This would be nothing more than yet another metaphysical, representational account still operating in terms of subjective consciousness and perception (even if for cultural groups instead of for individuals). In short, it lapses back into an even more derivative and inadequate conceptual scheme.

Rather, the hermeneutic point is that the manifestation of divinity in nature is a genuine accomplishment in the unfolding of earth, heavens, and humans, that is, a specific phase in the primal endowment of the world. This primal gift does appear in specific historical configurations, which, as we have seen, in their mode of disclosure are simultaneous with modes that necessarily are hidden. Nevertheless, even if the most astute thinkers and artists of the nineteenth century saw the character of what they were helping to bring about, they could not have been aware of that to which they were oblivious—how, in principle or in fact, could they have been? Furthermore, they could not have seen that, beyond the revelation and concealment happening in their lives and times, the manner in which the world happens (the primal way in which divinities, humans, heavens, and earth are meted out into existence and in relation with one another) itself hides the original event of this granting of world, the still-coming, primal granting-gathering of dimensions of world ("*Ereignis*," Heidegger often calls it).[93]

They could not have understood the God of the Old and New Testaments as a configuration of this giving, since they saw him as the creator of all creation, as the giver of all gifts, as prior to and independent of time and space. They could not have seen nature as one configuration of sky and heavens, since they saw it as creation, that is, as already formed and as already given to humans in the specific manner recounted in Genesis. How could they possibly have seen that in their own disclosed world, nature itself was something else more deeply hidden and only partially revealed, and even then revealed only in a way that forsook its own inherent nature? Since they positively participated in the disclosure of nature as divine symbol and site for human moral action (common to both versions of Genesis), they could not see that this manifestation of nature deforms it in its subjugation to representational concepts, where it can appear as having meaning only as a symbol and value, only as a counter between the spiritual action of God and his blessed American faithful.

Given that this concept was so enslaved and deformed, any other dimensions of earth and heavens that might have appeared—any portrayals of nature as other than landscape—were unimaginable not only because theological and artistic sensitivity and perception could not fully articulate them but also because the hidden mystery could not possibly occur to the imagination of these people in this time and place. To put it, not entirely appropriately, in the anthropomorphic terms of the Old Testament, it is as if the earth and heavens of America (itself yet another concept) were as the Israelites held in captivity: as long as they were captive they could not come into their own, destined nature; they could not come to fulfillment while in the hands and cultural constructs of their captors. Strangely, however, this exile from their own lands and essence also was a necessary part of their coming into their proper place. What are the possibilities for a captive who appears to his captor as nothing in himself, as merely a body or object to be disposed of according to the will of his master, with no hope of escape? Such a one might resist and struggle before being overpowered. He might resignedly go along and be considered a collaborator or prostitute. He might become a kind of curiosity, to be shown off to visitors and tourists.

So too with the earth and heavens. Under the reign of natural divinity they appeared as objects to be seen, used, enjoyed, and beguiled or forced to appear in artistic and technological projects. What could happen except that earth and heavens would appear as nature, as representational symbol and object? While bound up and seen only according to these concepts, as not possibly intelligible in terms other than "nature," "nature as landscape," or "the landscape of natural divinity"—all of which are versions of one and the same event—earth and heavens could not come to their own nature as inherently valuable and meaningful, yet still with a profound relation to humans and, perhaps, to divinity. Where earth and heavens could occur as the effect of the Judeo-Christian God or, alternatively, as the merely physical evolution of materiality, they could not come into their proper essence according to a yet more primal measure, in a giving and taking where earth and heavens are gathered and scattered.[94] Such an occurrence likely would disclose earth and heavens in a mode that we might approximate by the names *holy* or *mystery*—if we could say and think them nonmetaphysically.

There currently is a good deal of talk about the earth as a total, self-sustaining ecosystem, the so-called Gaia hypothesis. This shows our urge to account for nature in some sort of spiritual manner, in terms of a kind of wholeness to which we can belong and from and toward which we can responsibly act. Of course, to see the earth this way is substantially (although perhaps in a new phase) to translate it into the conceptual apparatus of systems logic and process philosophy. To say this is not to discount the movement but to describe it as still within philosophical and scientific representational thinking. Perhaps here we catch a glimpse of the next epoch of the manifestation of what was nature, a hint of the next historical era. In any case, however, such *systemic thinking*, even if it would achieve a posthumanistic, postanthropological attitude to the earth and its life, perhaps bringing about some new good, still would not attain the nonmetaphysical realm that hermeneutics seeks. Given the yearning for something beyond the metaphysical tradition of human domination over nature that has been sanctioned by the authority initially derived from God and now held by science, deep ecology not surprisingly emerges and seeks to find its way, partially by using Heidegger's thought.[95] What will work out here remains to be seen.

In any case, the nonmetaphysical opening up of earth and heavens has not yet been given. What ultimately was hidden from the nineteenth century by the disclosure of natural divinity and what still is hidden from us today by the modes of appearance of our own technological world is this event of earth and heavens as mystery, an event that may or may not yet occur and thus a realm in which we may or may not be able to dwell.[96]

Postscript

Postscript

As noted in the introduction, the issues surrounding the new theories and practices of interpretation are interesting and important. Fierce intellectual and cultural feuds are underway, the outcome of which will affect our culturally sanctioned ways of understanding the world and the range of individual differences tolerated by research funding and in publishing contracts and relevant to the success and abridgment of careers. The stakes are too high to remain uninformed or without a position. We all must participate in the decisions about social assumptions and attitudes and about the possibilities for academic and applied professional work.

This task obviously calls for additional attention to the difficult theories themselves. Ultimately, however, each of us must attempt to apply the alternatives in his or her own areas of interest. To provide initial help in that project, in chapter 1, I presented Wittgenstein's and Jung's views as paradigms of the major possibilities and choices: respectively, to move on by sweeping away misleading past assumptions and connotations in a process that never will be completed or to recover and transform complex, deep symbolic meanings in a manner appropriate to our new modes of building and living.[1] In chapter 2, I deconstructed not only pyramids but the "official" readings of pyramids by such scholars as Giedion and Etlin.[2] That is, I intentionally opened up a site to which readers may

117

return if they wish. They can reread Giedion and Etlin and more fully compare these traditional approaches with the newly proposed deconstruction.

Similarly, the hermeneutic interpretation of American nature in chapter 3 is meant to be an example with multiple agendas. It is intended to show how hermeneutic retrieval would proceed and also to challenge pointedly the traditional and deconstructive methods. If readers choose to pursue the issues, beyond understanding what hermeneutics would do, they can work through the other contested readings of American nature. So, as a final example or pointer, and nothing more than that, what follows—written from the perspective of hermeneutics, since that is where I left off—is an outline of how readers might continue thinking through hermeneutics versus deconstruction versus traditional approaches.

Hermeneutics assumes a specific attitude toward phenomena, an attitude that Heidegger, Gadamer, and others developed to return attention to the ordinary world and everyday life. Unlike deconstruction, hermeneutic investigation does not produce a new kind of work or different-looking designs and plans. Hermeneutics is not likely to be turned into a new style or to shift attention to the exotic. Rather, as Heidegger insisted, we need to reflect on what is nearest, on what is so close that we do not see or think it.[3] It is so near that it does not occur to us to attend to it, nor would it be easy to do so, since it may be the means by which anything else, everything else, is given to us at all. Thus, what is nearest of all hides itself as the invisible context within which the focally given appears. That is why Heidegger worked so hard to recover the meaning of what was close by: beings, language, things, and Being.

Each person and group needs to try to discover, to listen to and see, what is nearest and hidden. For contemporary Americans the phenomena appropriate for hermeneutics include things such as the landscape around us. Our landscape is so close that we rarely notice it, because we instead perceive objects, events, and ourselves in it. Our landscape is ordinary. We live in it, photograph it, build subdivisions on it, and vow to save its mountain streams from pollution. To do a hermeneutics of the American landscape means, as we have begun to understand, to retrieve past meanings effaced by time and forgetfulness and obscured by historical shifts and changes

118

through which we act and see differently than we had before. Hermeneutics attempts to peer into and through such erasures and disclose what was unthought by tracing phenomena back to their original meanings, back to the source that still comes to us and informs our culture and possibilities today. What had been taken for granted might be newly known and, in turn, transform and enrich our experiences and actions.

It will be objected, of course, that the American landscape is far from ignored. It is the focus of popular attention in constant media coverage of environmental issues and ecological debates. It is the theme of numerous artistic and scholarly projects.

It is precisely here, however, that we can see why hermeneutics is appropriate. Insofar as nature is seen as an object or problem to be dealt with by way of the latest technology or, subjectively, in terms of artistic "creativity," we still are within the metaphysical sphere. The continuing grasp of modernity is obvious all around us: the soft-focus, all but pastel marketing images generated by chemical companies present their creators as friends of the earth; subdivisions promise "Lake-Hills Meadows" or "River-Cloud Place"; lawn mower advertisements trim up our public face; nostalgic paintings of Native Americans in wilderness settings sell for thousands of dollars; and New Age crystals promise to bring cosmic harmony. All these promote, depend on, and in their own ways obscure the deep and originary meaning of our landscape. Have we learned nothing from Wittgenstein, Jung, Derrida, Gadamer, Eliade, and Heidegger, who all argue that we must strip away the illusion that we know what is about us, that what is here is obvious and unproblematic?

Other options remain, however. A direct comparison of hermeneutics with deconstruction, critical theory, and traditional scholarship can be nicely made by considering the recent trajectories of scholars pushing the critical investigation of the social, cultural, economic, and ideological context of nineteenth-century American painting. In the most controversial and extreme case, the posture of American art and the landscape of democracy have been partially deconstructed in the widely reviewed 1991 traveling exhibition and catalogue *The West as America: Reinterpreting Images of the Frontier, 1820–1920*.[4] Here, although deconstructive theory (strangely) is not mentioned, the grounding assumptions and traditional readings are rejected in inversions that bring the previously marginalized to the fore.

The subsequent 1994 exhibition *Thomas Cole: Landscape into History* and the book of the same title[5] shift from deconstruction to utilize the tactics of critical theory and cultural studies and are perhaps more likely permanently to change the course of the interpretation of Cole and American painting generally. William H. Truettner sets the tone by skewering traditional scholarship and its guiding metaphor:

> Scholars and collectors in the 1930s believed [Cole's] landscapes more or less truthfully represented nature. These same landscapes might also convey strong personal feelings, but they corresponded to what the artist had felt when observing a particular scene. Despite occasional distortions, Cole provided what was then called a window on the past—the look, the spirit, the unadorned beauty of the American wilderness in the second quarter of the nineteenth century.
>
> In addition, Cole seemed to provide that window without a lot of artistic fuss. The style of his landscapes, scholars argued, was nature's own—simple, direct, the product of a democratic culture in which academic art had in many instances emerged from folk art.[6]

In contrast, the exhibition and catalogue are meant to demonstrate that "scholars no longer need to focus on how Cole reproduced the look of wild nature, but on how he used both landscapes and history paintings as inventions to address the complex relationship between society and nature during his own era."[7]

The shift away from considering painting as a traditional, objective representation or as a source of deeper, hermeneutic meaning, it is argued, can be accomplished by viewing Cole in the light of the divided society of the Jacksonian era and its "struggle with complex social, political, and cultural issues—the rise of Jacksonian democracy, economic expansion, and the early phases of industrialization."[8]

Specifically, Alan Wallach's essay "Thomas Cole: Landscape and the Course of Empire" considers the tensions between artist and patron during the beginning stages of the market for landscape paintings in America and then as the interests and aesthetic preferences of an aristocratic elite yielded to those of a bourgeois or middle-class elite.[9] Although the artist participates in the patrons' competition for status, he or she also serves their interests by helping to accomplish and legitimize the commodification and consumption of the landscape. Wallach writes, "To depict a nature 'new to Art' constituted an act of appropriation. Nature had to be seized, tamed, brought under the dominion of artistic law."[10]

In a parallel analysis, "The Advantages of Genius and Virtue: Thomas Cole's Influence, 1848–58," J. Gray Sweeney probes the "artistic and social agendas" that are served as Cole's influences and critical reception are played out.[11] The rivalry for visibility, reputation, and sales is crucial, he argues, to understanding how Frederic Edwin Church, Asher B. Durand, Jasper F. Cropsey, Sanford R. Gifford, John F. Kensett, and many others deployed themselves in relation to Cole's influence and toward their critics and patrons. In any case, the consideration of status, power, and money at least implies that both traditional and conservative hermeneutic interpretations are too simple and economically and ideologically naïve.

To make a final comparison, consider part of the Columbian Quincentenary, the exhibition and catalogue *Columbus of the Woods: Daniel Boone and the Typology of Manifest Destiny*, in which J. Gray Sweeney revises the conventional views of Daniel Boone in the light of an analysis of cultural and ideological drives to power.[12] Sweeney holds that the artistic representations of Boone are linked to contending agendas for motivating and justifying expansion: the image of Boone is successively constituted so as to legitimize early land appropriation and speculation; to warn against the dangers of egalitarian tendencies and the ambitions of the lower classes; to "realistically" portray the actions of common individuals as being in the service of civilization; to symbolize heroic, divinely ordained empire building; and to indicate prophecy fulfilled as the lands of the Far West were assumed.

Obviously, these sources interpreting the landscape and painting allow direct comparison with chapter 1's traditional, biographical-cultural approach and with chapter 3's hermeneutics. The socioeconomic disassemblies using deconstruction and critical theory, although deft at taking the cultural forms apart, do not posit objective, "real" meanings, as the tradition holds, nor do they seek to fuse horizons or positively recover lost or hidden meanings, as does hermeneutics.

Readers interested in pursuing deconstruction and related approaches such as critical theory and varieties of postmodernism and poststructuralism can explore the booming literature that is reinterpreting the environment. Critical social theory and versions of late Marxism and post-Marxism, perhaps at their best in the work of scholars such as Denis Cosgrove, examine in detail how the landscape is a cultural construct. In his *Social Formation and Symbolic Landscape* and *The Iconography of Landscape,* Cosgrove and his colleagues explore landscape iconography within a broad social and economic context, especially analyzing how material appropriation of the land occurs by way of technical practices such as land surveys and reclamation, in painting and mapping, and in philosophical and literary discourse.[13] Other geographers and cultural theorists blend these theories and postmodernism to analyze the U.S. urban scene. Edward W. Soja and Michael Dear focus on Los Angeles, respectively, in *Postmodern Geographies* and "Taking Los Angeles Seriously: Time and Space in the Postmodern City"; Manuel Castells explores the technological urban landscape in *The Informational City.*[14]

Poststructural, rather than deconstructive, forays into the scene are undertaken by major French theorists: Baudrillard records two volumes of fracturing snapshots of his experiences traveling across the United States, and Lyotard imaginatively reappropriates and applies the sublime.[15] This approach to landscape as manufactured and utilized within systems of social and economic control (class struggle) and ideology also is variously developed in Ann Bermingham's *Landscape and Ideology,* Patricia Limerick's *Legacy of Conquest,* Alexander Wilson's *Culture of Nature: North American Landscape from Disney to the Exxon Valdez,* John D. Dorst's *Written Suburb,* Sharon Zukin's *Landscapes of Power,* Randall H. McGuire and Robert Paynter's *Archaeology of Inequality,* Stephen Greenblatt's *Marvelous Possessions,* and W. J. T. Mitchell's *Landscape and Power.*[16]

My main point, of course, is that all of us need to continue questioning the approaches and most importantly, through them, the subject matter. Although hermeneutics and deconstruction are more "newsworthy" at the moment, because they are battling about whether there is any meaning beyond the release that comes from the disassembly of oppressive, uncritically held constructions, those who trust a commonsensical understanding of objective "facts" and causality continue to develop traditional scholarship. Recent historical analyses in cultural geography such as those found in the essays in Michael Conzen's edited collection *The Making of the American Landscape* convey an appreciation of the diversity and complexity behind the heterogeneous aspects of the American landscape. Such work is more than correct; it largely succeeds in its ambitious goal: "The book aims at an unabashedly evolutionary interpretation of the American Landscape."[17] Parallel investigation is undertaken for literature in David Wyatt's *The Fall into Eden* and for visual documents by Anne Hyde in *An American Vision*.[18] Surprisingly, the traditional approach has not yet mustered a full theoretical (as opposed to a defensive political) response to the challenges from deconstruction and other poststructuralist positions. Surely it will rise to the occasion.

To keep the question open and the outcome moving along, from the hermeneutic point of view it would be argued that, at the moment at least, this mainstream work still tends to see the landscape through unquestioned rationalistic concepts, for example, by way of diffusion of forms or styles that materially represent cultural norms and patterns. As Conzen claims, the landscape indeed is thought in terms of one continuous evolutionary lineage. Heidegger would counter that this work, although helpful and correct in its results, nonetheless remains within the conventions of metaphysical representation and thus cannot question, much less break through, undo, and recover, the layers of meaning to which it itself contributes.[19] As Foucault would argue, the traditional approach overlooks the flux of *discontinuities*—the very stuff of which our rich cultural world is made.[20]

From the position of hermeneutics and deconstruction (remember, the tradition would have a rebuttal), a striking example of the limitation of unquestionably first-rate scholarship can be found in the examination of the so-called Hudson River school of painters in relation to the American landscape. Recently this phenomenon has been named, even definitively fixed, with the term *luminism* in, for instance, *American Light: The Luminist Movement,* edited by John Wilmerding in 1986, and Barbara Novak's *Nature and Culture.*[21] Although these interpretations do establish categorical identities and differences, they nonetheless founder by virtue of their own achievement. The definition of luminism was useful in locating and promoting an aspect of American landscape painting, but it has not been genuinely believed or taken up, even though it now informs the canon of studies. As Derrida and others would point out, it has posed as a new authoritative structure, not adequately self-critical about its own fictive character or its purpose within art history and cultural revaluation.

At the same time, luminism as a formal category has no authentic historical foundation. The term did not come into use until after the period to which it is retroactively applied. Although such a backward insertion could be part of either a deconstructive tactic to undercut the historical accumulation of validity or a hermeneutic experience through a retrieval of what was previously concealed, that certainly is not what these traditional scholars intend. Again we see the straightforward methodology of modern scholarship: classify and arrange according to formal elements, concepts, and categories and then look for and describe apparently unproblematic causes and diffusion patterns. This work, no matter how meticulous, no matter how widely adopted by the discipline, does not clear away its own apparatus and assumptions to recover its own origin in a historically hidden source.

Another approach may be necessary if we are to understand the landscape in a vital manner, experiencing new, "surplus" meaning by fusing our context, "prejudices," and concerns with earlier ones. To retrieve something of the still-vital source that came as both disclosed and hidden in the constitution of the nineteenth-century landscape and world, a self-critical, even radical environmental hermeneutics might be required. A hermeneutics of midnineteenth-century traditions of iconography and style would be a first step in discovering what landscape paintings and projects, with their emblematic and typological symbolism, have to show us today about possible attitudes toward nature and responsible, appropriate courses of environmental action.

Despite all our changes, we still react against and within the complex of meanings that nineteenth-century artists wove in connection with nature, when they helped to build our current home out of the American wilderness. Their views, in fact, are in many respects more sophisticated than those of today's scholars who operate—without knowing it—with the last in the chain of modernity's derivative, abstract concepts such as "objective homogeneous space" and "subjective artistic representation" and thus typically believe that "painters convert space, which has neither cultural nor personal associations, into familiar places."[22] That is why the aim of chapter 3 was to begin to work out a fundamental understanding of the American landscape by attending to the way in which artists interpreted it in their paintings during the nineteenth century and how these interpretations continue to bear on us.

This postscript's prodding to "take on" the other two approaches in regard to subject matter that really counts is written in the voice of hermeneutics because that is how chapter 3 speaks. The other two approaches would speak differently in making their own presentations. To encourage readers to listen to all three, I have acknowledged their spokespersons. We can imagine them waiting in the wings, wanting equal attention from the readers as the debate goes on—well beyond the realm that we have shared through this book. I hope that readers will listen to all these other voices, and will add their own, in thinking through phenomena such as our houses, public buildings and spaces, and the landscape, and in deciding on the worth of all three alternative modes of environmental interpretation.

Notes

Introduction

1. The following comments on diffusion patterns apply only to the environmental disciplines. Heidegger's most widely read work in the environmental disciplines has been the collection of essays translated by Albert Hofstadter as *Poetry, Language, Thought* (New York: Harper and Row, 1971), which includes "The Origin of the Work of Art," "Building Dwelling Thinking," "The Thing," and "...Poetically Man Dwells." *On The Way to Language,* trans. Peter Hertz and Joan Stambaugh (New York: Harper and Row, 1971), also has been influential, as, of course, has *Being and Time,* trans. John Robinson and John Macquarrie (New York: Harper and Row, 1962). The two works of Gadamer with the biggest impact are *Truth and Method,* trans. G. Barden and J. Cumming (New York: Seabury, 1975; the revised edition and new translation by Joel Weinsheimer and David G. Marshall [New York: Continuum, 1989] is a substantial improvement), and *Philosophical Hermeneutics,* trans. David Linge (Berkeley: University of California, 1976). Foucault became well known through such volumes as *Madness and Civilization,* trans. Richard Howard (New York: Vintage, 1967); *The Archaeology of Knowledge,* trans. Alan Sheridan Smith (New York: Pantheon, 1972); *The Order of Things,* trans. Alan Sheridan Smith (New York: Random House, 1973); and *Discipline and Punish,* trans. Alan Sheridan Smith (New York: Vintage, 1979). Derrida became

a general influence when Gayatri Spivak's translation of *Of Grammatology* appeared (Baltimore: Johns Hopkins University Press, 1976), followed by a flood of work published by the University of Chicago Press, including *Spurs: Nietzsche's Styles,* trans. Barbara Harlow (Chicago: University of Chicago Press, 1978); *Writing and Difference,* trans. Alan Bass (Chicago: University of Chicago Press, 1978); *Positions,* trans. Alan Bass (Chicago: University of Chicago Press, 1981); *Dissemination,* trans. Barbara Johnson (Chicago: University of Chicago Press, 1981); and *Margins of Philosophy,* trans. Alan Bass (Chicago: University of Chicago Press, 1982).

2. Although new ideas usually first appear in conference presentations and journal articles, widespread readership by members of other disciplines normally occurs when articles are included in collections. The following are some readily available books presenting the new methodologies. François Lyotard's *Postmodern Condition: A Report on Knowledge,* trans. G. Bennington and B. Massumi (Minneapolis: University of Minnesota Press), appeared in 1984, as did *Driftworks,* trans. Roger McKeon et al. (New York: Semiotext(e)). Baudrillard's *Mirror of Production,* trans. Mark Poster (St. Louis: Telos), was published in English in 1975 but became more popular and widely distributed in the 1980s after the appearance of works such as *Simulations,* trans. Paul Foss, Paul Patton, and Philip Beitchman (New York: Semiotext(e), 1983), and *America,* trans. Chris Turner (New York: Verso, 1988). Although both Gilles Deleuze and Félix Guattari—like most of these theorists—had established philosophical careers before their postmodern popularity, they became widely read through books such as *A Thousand Plateaus: Capitalism and Schizophrenia* (Minneapolis: University of Minnesota Press, 1988). De Certeau published *The Practice of Everyday Life,* trans. Steven Randall (Berkeley: University of California Press), in 1984 and *Heterologies: Discourse on the Other,* trans. Brian Massumi (Minneapolis: University of Minnesota Press), in 1989. Luce Irigaray's *Speculum of the Other Woman,* trans. Gillian C. Gill (Ithaca, N.Y.: Cornell University Press), first came out in 1985; Julia Kristeva's work in *The Kristeva Reader,* ed. Toril Moi and trans. Léon Roudica, Séan Handy, et al. (Oxford: Basil Blackwell), appeared in 1987. Among the many works by Jürgen Habermas and Alasdair MacIntyre, see Habermas, *Knowledge and Human Interests,* trans. Jeremy Shapiro (Boston: Beacon, 1971), and *The Theory of Communicative Action,* 2

vols., trans. Thomas McCarthy (Boston: Beacon, 1984, 1987); MacIntyre, *Whose Justice? Which Rationality?* (Notre Dame, Ind.: University of Notre Dame Press, 1988) and *Three Rival Versions of Moral Inquiry* (Notre Dame, Ind.: University of Notre Dame Press, 1990).

3. Walter Benjamin, *Reflections: Essays, Aphorisms, Autobiographical Writings,* ed. Peter Demetz and trans. Edmund Jephcott (New York: Schoken, 1978). Benjamin's unfinished project, *Das Passagen Werk* (The Arcades project, 1982), ed. Rolf Tiedemann, is vol. 4 of his *Gesammelte Schriften,* ed. R. Tiedemann and Hermann Schweppenhäuser, 6 vols. (Frankfurt am Main: Suhrkamp, 1972–). The project came to the attention of a wider English-speaking audience in 1989 with the work of Susan Buck-Morss, *The Dialectics of Seeing: Walter Benjamin and the Arcades Project* (Cambridge, Mass.: MIT Press, 1989). Henri Lefebvre, *Everyday Life in the Modern World,* trans. Sacha Rabinovitch (Harmondsworth, England: Penguin, 1971); Lefebvre, *The Production of Space,* trans. Donald Nicholson-Smith (Oxford: Blackwell, 1991). Fredric Jameson, "Architecture and the Critique of Ideology," in *The Ideologies of Theory: Essays 1971–1986,* 2 vols. (Minneapolis: University of Minnesota Press, 1988), 2:35–60. David Harvey, *The Condition of Postmodernity: An Enquiry into the Origins of Cultural Change* (Oxford: Blackwell, 1989). Edward Soja, *Postmodern Geographies: The Reassertion of Space in Critical Social Theory* (New York: Verso, 1989). Denis E. Cosgrove, *Social Formation and Symbolic Landscape* (Totowa, N.J.: Barnes and Noble, 1984); Denis E. Cosgrove and Stephen Daniels, eds., *The Iconography of Landscape* (New York: Cambridge University Press, 1988).

4. Christian Norberg-Schulz, *Genius Loci: Towards a Phenomenology of Architecture* (New York: Rizzoli, 1979); Norberg-Schulz, "Heidegger's Thinking on Architecture," *Perspecta* 20 (1983): 61–68; Norberg-Schulz, *The Concept of Dwelling: On the Way to Figurative Architecture* (New York: Rizzoli, 1985). Karsten Harries, "Thoughts on a Non-Arbitrary Architecture," *Perspecta* 20 (1983): 9–20, and "Space, Place, and Ethos: Reflections on the Ethical Function of Architecture," *Artibus et Historiae* 9 (1984): 159–165. Anne Buttimer's "Home, Reach, and Sense of Place," in *The Human Experience of Space and Place,* ed. A. Buttimer and D. Seamon (London: Croom Helm, 1980), 166–187, was widely circulated (the entire volume edited by Buttimer and Seamon was widely influential); David Seamon, *A Geography of*

the Lifeworld (London: Croom Helm, 1979); Seamon and Robert Mugerauer, *Dwelling, Place and Environment* (Dordrecht, Holland: Nijhoff, 1985; reprint, New York: Columbia University Press, Morningstar Editions, 1989). Edward Relph, *Place and Placelessness* (London: Pion, 1976) and *Rational Landscapes and Humanistic Geography* (London: Croom Helm, 1981).

5. For example, phenomenology can be seen as a basic component of, and variation on, hermeneutics, or it can be taken as the continuation of the realist tradition that stretches from Aristotle, through Aquinas, to Brentano. I have written extensively about the relation of phenomenology to hermeneutics elsewhere: Mugerauer, "Phenomenology and Vernacular Architecture" in *Encyclopedia of World Vernacular Architecture,* ed. Paul Oliver, 4 vols. (London: Blackwell, forthcoming), and "Phenomenology and the Environmental Disciplines," in the University of Texas Community and Regional Planning Program Working Paper Series (Austin: University of Texas, 1992). As noted previously, abundant examples of the phenomenology of the built environment are available. See, in addition to the works cited in note 4, my "Architecture as Properly Useful Opening," in *Ethics and Danger: Essays on Heidegger and Continental Thought,* ed. Charles Scott and R. Dallery (Albany, N.Y.: SUNY Press, 1992), 215–226; Mugerauer, "Toward a Phenomenology of Hot and Humid Climates," in *Architecture in Hot and Humid Climates,* ed. Wayne Attoe, forthcoming; and Mugerauer, "Toward an Architectural Vocabulary: The Porch as Between," in *Dwelling, Seeing, and Designing: Toward a Phenomenological Ecology,* ed. David Seamon (Albany, N.Y.: SUNY Press, 1992), 215–226.

6. Here I refer to works that take up and use a specifically deconstructive or hermeneutic approach to the built environment. In addition to consulting scattered articles in *Assemblage* and *Threshold,* see especially Mark Taylor's "Architecture of Pyramids," *Assemblage* 5 (Feb. 1988): 17–27, and "Deadlines: Approaching an Architecture," *Threshold* 4 (Spring 1988): 20-27, although the former is more textually oriented than the latter. Denis Hollier's *Against Architecture: The Writings of George Bataille* (Cambridge, Mass.: MIT Press, 1989) is similarly textual in emphasis and not as direct for architects as the title might indicate.

Two of the most promising publications are Peter Jukes's *A Shout in The Street: An Excursion into the Modern City* (Berkeley: University of

California Press, 1991) and Dennis Crow's introduction to the collection he edited entitled *Philosophical Streets* (Washington, D.C.: Maisonneuve, 1990), 1–26. Paul Virilio covers architecture and the city in *Lost Dimension* (New York: Semiotext(e), 1991), but the work is more a creative, poststructural rereading than a disciplined heuristic deconstruction.

More explicitly deconstructive are Michael Benedikt, *Deconstructing the Kimbell* (New York: SITES/Lumen, 1991); Bonnie Bridges and Robert Mugerauer, "Recasting the Body Politic: Deconstructing the Athenian Agora," in *Bodies: Image, Writing, Technology,* ed. Juliet MacCannell and Laura Zakarin (Albany, N.Y.: SUNY Press, forthcoming); Robert Mugerauer, "Post-Structuralist Planning Theory," University of Texas Community and Regional Planning Program Working Paper Series (Austin: University of Texas, 1991). Hermeneutics is used in Mugerauer, "The Post-Structuralist Sublime: From Heterotopia to Dwelling?" a talk presented in Jan. 1991 at the University of Minnesota School of Architecture and Landscape Architecture, where it was videotaped, and in Feb. 1992 at the University of Washington.

7. Heidegger discusses the Greek temple in "The Origin of the Work of Art," in *Poetry, Language, Thought,* 17–87. His discussion of the farmhouse as gathering the fourfold realm occurs in "Building Dwelling Thinking," which appears in the same volume, 154–161. Derrida's initial forays into architecture were relatively obscure, as with his interview in *Domus* 671 (April 1986): 17–24. Work in book form was more available, such as "The Pit and the Pyramid: Introduction to Hegel's Semiology," in *Margins of Philosophy,* 69–108. Lately his comments appear frequently in architectural publications, for instance, "In Discussion with Christopher Norris," in *Deconstruction: The Omnibus Volume* (New York: Rizzoli, 1989), 71–75; "Why Peter Eisenman Writes Such Good Books," *Threshold* 4 (Spring, 1988): 99–105; and the exchange of letters with Eisenman published in *Assemblage* 12 (1990): Derrida's "A Letter to Peter Eisenman," 7–13, and Eisenman's "Post/El Cards: A Reply to Jacques Derrida," 14–17. The record of the six sessions between Eisenman and his design group and Derrida on the *"chora"* project will appear shortly in Jeffrey Kipnis, *Choral Work* (London: Architectural Association, forthcoming).

8. Given the vast literature of twenty-five hundred years of theory, I despair to provide a bibliography for traditional interpretation. Among

the most useful secondary sources are John Hospers, *Meaning and Truth in the Arts* (Chapel Hill: University of North Carolina Press, 1946); Titus Burckhardt, *Sacred Art in East and West,* trans. Lord Northbourne (Bedford, England: Perennial, 1967); Amos Rapoport, *House Form and Culture* (Englewood Cliffs, N.J.: Prentice-Hall, 1969); Erwin Panofsky, *Meaning in the Visual Arts: Papers in and on Art History* (Garden City, N.Y.: Doubleday, 1955); Jacques Maritain, *Art and Scholasticism* (New York: Scribner's, 1962); Yvor Winters, *In Defense of Reason* (Denver: Alan Swallow, 1947); and *The Function of Criticism* (Denver: Alan Swallow, 1957). Of course, the basic sources, as noted, remain those of Plato, Aristotle, Hume, Kant, Hegel, and so on, which can be found through any standard reference work in aesthetics.

9. Winters, *In Defense of Reason,* 29.

10. Winters, *The Function of Criticism,* 26; Winters, *In Defense of Reason,* 17.

11. The major rereadings of Heidegger as more radical than generally has been appreciated are given by Reiner Schürmann, *Heidegger on Being and Acting: From Principles to Anarchy,* trans. Christine Marie Gros (Bloomington: Indiana University Press, 1987); John Caputo, *Radical Hermeneutics: Repetition, Deconstruction, and the Hermeneutic Project* (Bloomington: Indiana University Press, 1987); Gerald Bruns, *Heidegger's Estrangements: Language, Truth, and Poetry in the Later Writings* (New Haven, Conn.: Yale University Press, 1989).

12. Gadamer, *Truth and Method,* 263. The best secondary source on Gadamer's work is Joel Weinsheimer, *Gadamer's Hermeneutics: A Reading of Truth and Method* (New Haven, Conn.: Yale University Press, 1985). It is important to note, in addition to the influential works of Gadamer cited in note 1 (which are the main source of the version of hermeneutics presented here), the work of Paul Ricoeur, especially *The Conflict of Interpretations,* trans. Willis Domingo, et al. (Evanston, Ill.: Northwestern University Press, 1974), and *Hermeneutics and the Human Sciences,* trans. John B. Thompson (New York: Cambridge University Press, 1981); an especially useful secondary source is Don Ihde, *Hermeneutic Phenomenology: The Philosophy of Paul Ricoeur* (Evanston, Ill.: Northwestern University Press, 1971).

13. Richard E. Palmer, *Hermeneutics: Interpretation Theory in Schleiermacher, Dilthey, Heidegger, and Gadamer* (Evanston, Ill.: Northwestern University Press, 1969), 178. Palmer's book remains one

of the most useful secondary sources on hermeneutics, especially in regard to the background in intellectual history.

14. This paragraph draws its ideas from several sources. In order, they are Gadamer, *Truth and Method,* 375; Weinsheimer, *Gadamer's Hermeneutics,* 225; Gadamer, *Truth and Method,* 238.

15. Gadamer, *Philosophical Hermeneutics,* 121.

16. Gadamer, *Truth and Method,* 265.

17. Heidegger, *Being and Time,* 194–195; Gadamer, *Truth and Method,* 258–262; Gadamer, *Philosophical Hermeneutics,* 9, 18–42; and Gadamer's little-known but very accessible and useful "On the Circle of Understanding," in *Hermeneutics vs. Science? Three German Views,* ed. John M. Connolly and Thomas Keutner (Notre Dame, Ind.: University of Notre Dame Press, 1988), 68–78.

18. Gadamer, *Truth and Method,* 239.

19. Ibid., 259.

20. Ibid., 240–242.

21. Ibid., 244.

22. Ibid., 249.

23. Ibid., 246.

24. Ibid., 248.

25. Ibid., 245.

26. E. D. Hirsch, Jr., *Validity in Interpretation* (New Haven, Conn.: Yale University Press, 1967); Gadamer, *Truth and Method,* 264, 337; Weinsheimer, *Gadamer's Hermeneutics,* 156–157.

27. Gadamer, *Truth and Method,* 264.

28. Ibid., 253.

29. Ibid., 273.

30. Palmer, *Hermeneutics,* 180.

31. Ibid., 121.

32. Derrida, "Restitutions of the Truth in Pointing [*pointure*]," in *The Truth of Painting,* trans. Geoff Bennington and Ian McLeod (Chicago: University of Chicago Press, 1987), 255–382; quotation on 274.

33. Derrida, *Of Grammatology,* 158.

34. Northrop Frye, "Second Essay: Ethical Criticism: Theory of Symbols," in *Anatomy of Criticism* (New York: Atheneum, 1968), 71–128. Frye's seminal analysis of the five phases of the verbal symbol, in which the centripetal and centrifugal forces of words play out, remains an important key to understanding today's debates.

Specifically, the debate between traditional and deconstructive theory carries on the question of whether the final force of symbols is outward or inward.

35. Derrida, *Of Grammatology,* 61.

36. Derrida, *Disseminations,* 245–246.

37. Ibid., 250.

38. Ibid., 333–334.

39. Peter Dews, *Logics of Disintegration: Post-Structuralist Thought and the Claims of Critical Theory* (New York: Verso, 1987), 12–13.

40. Derrida, *Positions,* 45.

41. Dews, *Logics of Disintegration,* 13; Derrida, *Positions,* 45.

42. Derrida, "Differance," in *Speech and Phenomena,* trans. David B. Allison (Evanston, Ill.: Northwestern University Press, 1973), 129.

43. Dews, *Logics of Disintegration,* 11.

44. Derrida, *Spurs,* 57, 67; Caputo, *Radical Hermeneutics,* 151–152, 155–156.

45. Derrida, *Margins,* 123–126.

46. Ibid., 124.

47. Ibid., 125.

48. Derrida, *Spurs,* 97.

49. Ibid., 87–97.

50. Ibid., 107; compare with Derrida, *Dissemination,* 25.

51. Richard Harland, *Superstructuralism: The Philosophy of Structuralism and Post-Structuralism* (New York: Methuen, 1987), 135.

52. Derrida, *Dissemination,* 324.

53. Derrida, *Edmund Husserl's Origin of Geometry: An Introduction,* trans. John P. Leavey, Jr. (Stony Brook, N.Y.: Nicholas Hays, 1978), 88.

54. Derrida, *Writing and Difference,* 25.

55. Derrida, *Dissemination,* 328.

56. Derrida, *Of Grammatology,* 159.

57. Derrida, *Memoirs of the Blind: The Self-Portrait and Other Ruins,* trans. Pascale-Anne Brault and Michael Naas (Chicago: University of Chicago Press, 1993), 15.

58. Derrida, *Dissemination,* 324.

59. Derrida, *Writing and Difference,* 20.

60. Derrida, *Spurs,* 57; Caputo, *Radical Hermeneutics,* 156.

61. Derrida, *Of Grammatology,* 86.

62. Derrida, *Dissemination,* 328.

63. Derrida, *Of Grammatology*, 86.

64. "What I call the erasure of concepts ought to mark the places of that future meditation. For example, the value of the transcendental arche must make its necessity felt before letting itself be erased" (Derrida, *Of Grammatology*, 61).

65. On Wittgenstein and Jung presenting the two basic options for today, see the first and last sections of chapter 1. The arguments of the tradition and hermeneutics for the existence of deep meaning obviously correspond to versions of what Jung holds. In addition, it should be noted that though attention usually is focused on the relation of Derrida to the nineteenth-century philosopher Friedrich Nietzsche, which certainly is appropriate, Derrida also moves in Wittgenstein's steps, a realization that recent scholarship finally is beginning to take up; see Richard Rorty, *Contingency, Irony, and Solidarity* (Cambridge: Cambridge University Press, 1989) and *Philosophical Papers*, 2 vols. (Cambridge: Cambridge University Press, 1991). There has been an increasing interest in construing Wittgenstein's practices of removing misleading and dangerous elements of thinking and writing as "protodeconstructivist." One of the first treatments was Henry Staten, *Wittgenstein and Derrida* (Lincoln: University of Nebraska Press, 1984), and a recent one is Newton Garver and Seung-Chong Lee, *Derrida and Wittgenstein* (Philadelphia: Temple University Press, 1994).

Chapter 1: Traditional Approaches

1. Noted by Norman Malcolm in *Ludwig Wittgenstein: A Memoir* (New York: Oxford University Press, 1958) and M. O'C. Drury in "Conversations with Wittgenstein," in *Ludwig Wittgenstein: Personal Recollections*, ed. Rush Rhees (Totowa, N.J.: Rowman and Littlefield, 1981).

2. Ludwig Wittgenstein, *Tractatus Logico-Philosophicus*, trans. D. F. Pears and B. F. McGuinness (London: Routledge and Kegan Paul, 1963).

3. See O. K. Bouwsma, *Philosophical Essays* (Lincoln: University of Nebraska Press, 1965), and Timothy Binkley, *Wittgenstein's Language* (Ph.D. diss., University of Texas at Austin, 1970), 178 (this work also has been published by Martinus Nijhoff, The Hague, in 1973 with the same title; page numbers here refer to the dissertation). For example, Wittgenstein said, "A *picture* held us captive. And we could not get out-

side it, for it lay in our language and language seemed to repeat it to us inexplicably" (*Philosophical Investigations,* trans. G. E. M. Anscomb [New York: Macmillan, 1965], §115). Wittgenstein, then, also held that we can learn from our errors and mistakes—an interesting point of comparison with Freud.

4. Binkley, *Wittgenstein's Language,* 181.

5. Wittgenstein, *Tractatus,* sections 3.4–3.42.

6. Binkley, *Wittgenstein's Language,* 47.

7. Ibid., 14–15.

8. Ibid., 181. Wittgenstein's topological analogies, metaphors, and mappings are explored by Robert J. Ackerman in *Wittgenstein's City* (Amherst: University of Massachusetts Press, 1988), although he does not cite Binkley's work, which was published by Nijhoff fifteen years earlier.

9. Bernhard Leitner, *The Architecture of Ludwig Wittgenstein: A Documentation* (New York: New York University Press, 1976), 11. In addition to Leitner's pioneering work on the subject, an interesting book has recently appeared: Paul Wijdeveld's *Ludwig Wittgenstein, Architect* (Cambridge: MIT Press, 1994) appears to be a thorough documentation of the villa project, wonderfully illustrated. Another invaluable source is Michael Nedo and Michele Ranchetti, eds., *Wittgenstein—Sein Leben in Bildern und Texten* (Frankfurt am Main: Suhrkamp, 1983).

10. Hermine Wittgenstein, "My Brother Ludwig," trans. Bernhard Leitner, reprinted in Rhees, *Ludwig Wittgenstein,* 6.

11. C. H. von Wright, "A Biographical Sketch," in Malcolm, *Ludwig Wittgenstein,* 10–11. In fact, later, speaking of van der Null's difficulty in a bad architectural and cultural period, Wittgenstein proposed resisting the architectural "common currency": "Don't take comparability, but rather incomparability, as a matter of course" (Wittgenstein, *Culture and Value,* ed. C. H. von Wright and trans. Peter Winch [Oxford: Blackwell, 1980), entry of 1947–1948, 74e.

12. Wittgenstein, *Culture and Value,* entry of 1930, 6e.

13. Ibid., entry of 1936, 145e. Compare with Adolf Loos's views on the "removal of the surplus" in architecture and life; see Loos, *Spoken into the Void: Collected Essays 1897–1900,* trans. Jane O. Newman and John H. Smith (Cambridge, Mass.: MIT Press, 1982), and Benedetto Gravagnuolo, ed., *Adolf Loos: Theory and Works* (New York: Rizzoli, 1988).

14. Leitner, *The Architecture of Ludwig Wittgenstein,* 11.

15. O'C. Drury, "Conversations with Wittgenstein," 121.

16. Ibid.

17. Letter of 1928 to M. Weber and Co., quoted in Leitner, *The Architecture of Ludwig Wittgenstein,* 124.

18. Hermine Wittgenstein, quoted in Leitner, *The Architecture of Ludwig Wittgenstein,* 7–8.

19. Wittgenstein, *Culture and Value,* 1932–1934, 34e.

20. Hermine Wittgenstein, quoted in Leitner, *The Architecture of Ludwig Wittgenstein,* 23.

21. Wittgenstein, *Culture and Value,* 1947–1948, 69e. This view seems to differ from Loos's. See Loos, *Spoken into the Void,* and Gravagnuolo, *Adolf Loos: Theory and Works.*

22. Wittgenstein, *Culture and Value,* 1932–1934, 22e.

23. Ludwig Wittgenstein, *Philosophical Investigations,* §329.

24. Wittgenstein, *Culture and Value,* 1930, 3e.

25. Ibid.

26. C. H. von Wright was perhaps the first to draw a connection between Wittgenstein's philosophy and architecture, although he overlooks the dynamic of Wittgenstein's philosophy as activity and therapeutic remark: "The building [Wittgenstein House] is the work down to the smallest detail and is highly characteristic of its creator. It is free from all decoration and marked by a severe exactitude in measure and proportion. Its beauty is of the same simple and static kind that belongs to the sentences of the *Tractatus*" (quoted in Malcolm, *Ludwig Wittgenstein,* 10–11).

27. Binkley, *Wittgenstein's Language,* 169 ff.

28. Wittgenstein, *Culture and Value,* 1930, 7e.

29. Ibid., 1942, 42e.

30. Ibid., 1931, 16e.

31. Ibid., 1931, 22e.

32. Ibid., 1940, 38e; Wittgenstein notes Søren Kierkegaard on the "hothouse plant" analogy.

33. Hermine Wittgenstein, commenting on a painting of her brother, notes that "Ludwig's face appears to me in reality as being . . . gaunt and flat, his curly hair striving upwards much more and literally resembling flames, which seem to suit the intensity of his character" (quoted in Rhees, *Ludwig Wittgenstein,* 10).

34. C. G. Jung, *Memories, Dreams, Reflections,* ed. Aniela Jaffé (New York: Pantheon, 1973), 20–23. Although I had not read the essay at the time that I completed the first drafts of this essay, I must acknowledge that Clare Cooper seminally influenced behavior and environment research by introducing Jungian issues of the relation of house and self-identity in "The House as Symbol of Self," in *Designing for Human Behavior: Architecture and the Behavioral Sciences,* ed. Jon Lang, et al. (Stroudsburg, Pa.: Dowden, Hutchinson, & Ross, 1974).

35. Jung, *Memories,* 81.

36. Ibid., 82.

37. Ibid., 45, 57, 88–89, 234.

38. Ibid., 197–198.

39. Ibid., 199; for another example, see 213.

40. Ibid., 174–175.

41. Ibid., 160–165. In an interesting parallel to Jung's insight into the unfinished nature and unfolding relations among dreams, the psyche, and the image of the house, Wittgenstein appreciated the incremental and developmental character of language, thought, and life. Wittgenstein uses a strikingly similar figure: "Our language can be seen as an ancient city: a maze of little streets and squares, of old and new houses, and of houses with additions from previous periods; and this surrounded by a multitude of new boroughs with straight regular streets and uniform houses" (*Philosophical Investigations,* §18).

42. Jung, *Memories,* 134 ff.

43. Ibid., 138–139.

44. C. G. Jung, *The Archetypes of the Collective Unconscious,* vol. 9, pt. 1, of his *Collected Works,* 2d ed., ed. R. F. C. Hull, 20 vols. (Princeton, N.J.: Princeton University Press, 1967–1979), 275; cf. paragraph 266 in *Two Essays on Analytical Psychology,* vol. 7 of *Collected Works.*

45. Jung, *Memories,* 223.

46. Ibid., 225, 237.

47. Compare Jung's insight during a trip to Africa, related in *Memories,* 272–273.

48. Jung, *Memories,* 224.

49. Ibid.

50. Ibid.

51. Ibid., 225.

52. Ibid., 225–226.

53. Ibid., 225–229.
54. Ibid., 225.
55. Ibid., 237; cf. 225.

Chapter 2: Deconstruction

1. Derrida, *Husserl's Origin of Geometry,* 88. Derrida regularly uses the pyramid to think the difference between *difference* and *différance,* absence and presence, life and death. In treating the difference between speaking and writing and the silence and the voiced, he discusses the *e* in *difference* and the *a* in (the neologism) *différance:* "[Difference] is put forward by a silent mark, by a tacit monument, or, one might say, by a pyramid—keeping in mind not only the capital form of the printed letter but also that passage from Hegel's *Encyclopedia* where he compares the body of the sign to an Egyptian pyramid. The *a* of differance, therefore, is not heard; it remains silent, secret, and discreet, like a tomb" ("Differance," in *Speech and Phenomena* [Evanston, Ill.: Northwestern University Press, 1973], 132).

2. On difference, displacement, and deconstruction, see the discussion of Jacques Derrida in the introduction; see also Northrop Frye, "Theory of Symbols," in *Anatomy of Criticism,* especially 106. On the pyramid as sign and symbol for the thing itself, and the pyramid as text, see Derrida's reference to Hegel in "The Pit and the Pyramid" and the Derrida interview in *Domus.*

To undertake a deconstruction of the "official" or orthodox discourse that attempts to ground the pyramids is not to deny the power of this discourse, much less to question its "correctness" or scholarly achievement. On the contrary, its correctness and accomplishment are presupposed and accepted; it is precisely as determinative and authoritative that such discourse achieves the desired posture and carries out the necessary fictive strategies, which it simultaneously conceals by its very correctness and authority. Taking Sigfried Giedion's and Richard Etlin's accounts as cultural codifications (of the Egyptian and French neoclassic pyramids, respectively) is not to dispute or find fault with them but to accept and dislocate them in an attempt to disclose the posturing of the phenomena and of the sanctioning discourse and, thus, the deeper, concealed operations at work.

3. See Sigfried Giedion, *The Eternal Present: The Beginnings of Art* (New York: Bollingen Foundation, 1962); Mircea Eliade, *Patterns in Comparative Religion,* trans. Rosemary Sheed (New York: World, 1968), e.g. section 43; H. Frankfurt, *Kingship and the Gods* (Chicago: University of Chicago Press, 1948).

4. See Derrida, "The Pit and the Pyramid."

5. *Utterance 600,* trans. F. Mercer, quoted in Giedion, *The Eternal Present,* 275.

6. See Derrida, "The Pit and the Pyramid."

7. Alternatively, perhaps the pharaohs had so interiorized the belief in their identity with the sun god and in their own immortality and power that they saw the rest of the earthly realm as merely passing away, with nothing, not even predecessors' tombs, standing outside their own identity and presence. Here the "desecration" or reuse of materials would be a deeper incorporation into and assertion of their own presence.

8. The basic, straightforward account of the historical development of the French neoclassic tradition, which is used throughout and deconstructed in this essay, is taken from Richard A. Etlin, *The Architecture of Death: The Transformation of the Cemetery in Eighteenth-Century Paris* (Cambridge, Mass.: MIT Press, 1984).

9. On the sublime, see J. Gray Sweeney, *Themes in American Painting* (Grand Rapids, Mich.: Grand Rapids Art Museum, 1977), and chapter 3 of this book.

10. Hence the orthodox Catholic interpretation of Masonic belief as a "naturalistic religion." See note 24 on Masonry.

11. Etlin, *The Architecture of Death,* 108.

12. Ibid., 3.

13. On the organization of space and society at that time, also see Michel Foucault, *The Order of Things* and *Discipline and Punish.*

14. Etlin, *The Architecture of Death,* 17 ff.

15. Ibid., 62.

16. Ibid., 51, 55.

17. Ibid., 125.

18. The sublime and the seasons of life were popular themes for poetry and painting; see, for example, the later series by Thomas Cole, *The Voyage of Life,* explicated by J. Gray Sweeney in *Themes In*

American Painting and in *Natural Divinity,* a research project for the Smithsonian Institution's Museum of American Art.

19. Etlin, *The Architecture of Death,* 128.

20. Ibid., 146.

21. Ibid., 119.

22. Pace *ka.* Ibid., 128.

23. Ibid., 125.

24. Ibid.

25. Ibid. Etlin also notes that the triangle was a special, sacred symbol of the elements for the Freemasons, who also took the radiant triangle as a symbol of Jehovah, the great architect of the universe (ibid.).

26. See Foucault, *The Order of Things* and *Discipline and Punish;* Robert Mugerauer, "The Historical Dynamic of the American Landscape," paper presented to Council of Educators in Landscape Architecture, University of Illinois at Urbana, Sept. 1985; Mugerauer, *Interpretations on Behalf of Place* (Albany, N.Y.: SUNY Press, 1994).

27. Etlin, *The Architecture of Death,* 125.

28. Ibid.

29. Giedion, *The Eternal Present,* 72.

30. *Architect's International* 179, no. 7 (Feb. 15, 1984): 38.

31. This would bear out the influence of the "counterformal" aesthetic as developed by Robert Venturi in *Complexity and Contradiction in Architecture,* The Museum of Modern Art Papers on Architecture no. 1 (New York: Doubleday, 1966).

32. See Etlin, *The Architecture of Death;* Jacques François Blondel, *L'Architecture française (1752–56),* ed. Louis Savot (Geneva: Minkoff, 1973).

33. See Derrida, "The Pit and the Pyramid."

34. *Architect's International* 179, no. 7 (Feb. 15, 1984): 38.

35. John Pastier, "Isozaki's Design for MOCA," *Arts and Architecture* 2, no. 1 (1983): 31–34.

36. Ibid., 35.

37. See Martin Heidegger, *The Question Concerning Technology and Other Essays,* ed. and trans. William Lovitt (New York: Harper and Row, 1977); Robert Mugerauer, "From Technology to Dwelling," in *Interpretations on Behalf of Place,* 67–76.

38. See Richard Marx, "Egyptian Architecture in Los Angeles," *Los Angeles Times,* Sunday supplement, May 10, 1977.

39. Giedion, *The Eternal Present;* Eliade, *Patterns in Comparative Religion;* Frankfurt, *Kingship and the Gods.*

Chapter 3: Hermeneutical Retrieval

1. For the theoretical foundation of this analysis, from a Heideggerian approach, see Robert Mugerauer, "Language and the Emergence of Environment," in *Dwelling, Place and Environment,* ed. Seamon and Mugerauer, 51–70, and Reiner Schürmann's radical philosophical analysis of "historical economies" in *Heidegger on Being and Acting.* On Mircea Eliade's approach to an environmental hermeneutics, see Robert Mugerauer, *Interpretations on Behalf of Place,* chap. 4, 52–64.

A large part of this chapter is compatible with Gadamer's traditional approach. The art-historical interpretation implicitly closest to Gadamer's conservative hermeneutic probably is Robert Rosenblum's *Modern Painting and the Northern Romantic Tradition: Friedrich to Rothko* (New York: Harper and Row, 1975). A less conservative use of hermeneutics and Gadamer, closer to the approach of the last section of this chapter, is found in J. Gray Sweeney, *The Columbus of the Woods: Daniel Boone and the Typology of Manifest Destiny* (St. Louis: Washington University Gallery of Art, 1992), which I discuss in the postscript.

This chapter is especially concerned with the artistic use of natural and biblical typology to delineate a vision of nature's divinity. For treatment of the general issue, see Ursula Brumm, *American Thought and Religious Typology,* trans. John Hoaglund (New Brunswick, N.J.: Rutgers University Press, 1970); on typology in American rhetoric and literature, see Philip F. Gura, *The Wisdom of Words: Language, Theology, and Literature in the New England Renaissance* (Middletown, Conn.: Wesleyan University Press, 1981), and Sacvan Bercovitch, *The American Jeremiad* (Madison: University of Wisconsin Press, 1978) and *Typology and Early American Literature* (Boston: University of Massachusetts Press, 1971); on typology and the visual arts, see James Collins Moore, *The Storm and the Harvest* (Ph.D. diss., University of Indiana, 1974), and J. Gray Sweeney, *Natural Divinity,* unpublished project done as part of a Smithsonian Institute senior fellowship, 1984–1985. For a list of the basic monographs on the individual artists treated here, see the bibliographies in Barbara Novak, *Nature and Culture: American Landscape and Painting, 1825–1875* (New York: Oxford University Press, 1980); Joseph D. Ketner II and Michael J. Tammenga, *The Beautiful, the Sublime, and the Picturesque* (St. Louis: Washington University, 1984); and David C. Huntington, *Art and the Excited Spirit: America in the Romantic Period* (Ann Arbor: University of Michigan Museum of Art, 1972).

2. Mircea Eliade, "Paradise and Utopia: Mythical Geography and Eschatology," in *The Quest: History and Meaning in Religion* (Chicago: University of Chicago Press, 1969), 88–111. On the background of the idea of America as paradise, also see Charles L. Sanford, *The Quest for Paradise* (Urbana: University of Illinois Press, 1969); George H. Williams, *Wilderness and Paradise in Christian Thought: From the Garden of Eden and the Sinai Desert to the American Frontier* (New York: Harper and Row, 1962).

3. Because the Aramaic word meaning "garden" is translated by the Greek word meaning "paradise" in the Greek version of the Bible, *paradise* became a traditional equivalent of *Garden of Eden.* The contemporary English of the Jerusalem Bible translates Gen. 2:15 as "Yahweh God took the man and settled him in the garden of Eden to cultivate and take care of it."

4. Quoted in Sanford, *The Quest for Paradise,* 40.

5. Eliade, "Paradise and Utopia," 91.

6. Ibid., 93.

7. Ibid., 91. On American typology and sacred history, especially its secularization in the idea of progress during the eighteenth century, see Bercovitch, *The American Jeremiad,* chap. 4, "The Typology of America's Mission," 93–130; William Clebsch, *From Sacred to Profane America: The Role of Religion in American History* (New York: Harper and Row, 1968). On Protestant theology, hermeneutics, and religious thought, see Hans W. Frei, *The Eclipse of Biblical Narrative: A Study in Eighteenth and Nineteenth Century Hermeneutics* (New Haven, Conn.: Yale University Press, 1974); Jerry Wayne Brown, *The Rise of Biblical Criticism in America* (Middletown, Conn.: Wesleyan University Press, 1969); James West Davidson, *The Logic of Millennial Thought: Eighteenth-Century New England* (New Haven, Conn.: Yale University Press, 1977). On the theory of secular typological interpretation, see Louis H. Mackey, "Notes toward a Definition of Philosophy," *Franciscan*

Studies 33, no. 11 (1973): 262–272, especially sect. 3; Northrop Frye, "Levels of Meaning in Literature," *Kenyon Review* (Spring 1950): 246–262.

8. Sanford, *Quest for Paradise,* 52 ff., and George H. Williams, *Wilderness and Paradise,* 65 ff.

9. Quoted in Eliade, "Paradise and Utopia," 93.

10. Sanford, *Quest for Paradise,* 111. For the maps, see Seymour I. Schwantz and Ralph E. Ehrenberg, *The Mapping of America* (New York: Abrams, 1980), 49, 60.

11. William Cronon, *Changes in the Land: Indians, Colonists, and the Ecology of New England* (New York: Hill and Wang, 1983), 25 ff.; Smith is quoted in Eliade, "Paradise and Utopia," 94.

12. Of course, there are other important cultural interpretations of nature, such as the picturesque and arcadian. I omit those here for the practical matter of length and because (1) these are themselves variations on the two (Genesis) archetypes, (2) they are incorporated into a more powerful tradition of the "sublime," and (3) the arcadian itself is secularized from Roman religion.

Also, it is crucial to keep in mind this essay's focus and specific use and treatment of landscape painting, lest either the central thesis or the significance of the paintings become exaggerated. The claim here is that a religious understanding, rooted in Genesis, originally was a dynamic factor in the nineteenth-century interpretation of American nature, although that has long been forgotten. The claim is *not* that the two creation accounts in Genesis were the only, or even dominant, factors in the religious views of the time or even that the entire religious sensibility was all that mattered in the nineteenth century, which would be nonsense. Nor do I suggest that recovering the influence of the two Genesis accounts is, by itself, adequate to interpret fully any of the paintings considered here. On the issue of nineteenth-century painting in its religious context, see J. Gray Sweeney, "The Advantages of Genius and Virtue: Thomas Cole's Influence, 1848–1858," in *Thomas Cole: Landscape into History,* ed. William Truettner and Alan Wallach (Washington, D.C.: National Museum of Art, Smithsonian Institution, 1994), 113–135.

Where the second and third sections of this chapter provide close explications of the paintings, they agree with J. Gray Sweeney's interpretations and at times use the text of his *Themes in American Painting.*

He graciously has allowed the use of this material from his out-of-print work without the tangle of quotation marks that would substantially impede the reader.

13. Quoted in Henry Nash Smith, *Virgin Land: The American Land as Symbol and Myth* (New York: Vintage, 1950), 236. See also R. W. B. Lewis, *The American Adam: Innocence, Tragedy, and Tradition in the Nineteenth Century* (Chicago: University of Chicago Press, 1955), and Eliade, "Paradise and Utopia," 100–101.

14. Francis S. Grubar, *William Ranney* (Washington, D.C.: The Corcoran Gallery of Art, 1962), 32.

15. An interesting analysis of the "taste for landscape paintings" enabling the careers of Cole and his colleagues that focuses on economic, political, and class relationships has just been provided by Alan Wallach, "Thomas Cole: Landscape and the Course of Empire," in *Thomas Cole,* ed. Truettner and Wallach, 23–112.

16. Quoted in John W. McCoubrey, ed., *American Art, 1700–1960* (Englewood Cliffs, N.J.: Prentice-Hall, 1965), 102.

17. This hermeneutic of the religious principle behind the phenomenon, of course, would be considered hopelessly simple and retrograde from the viewpoint of deconstruction and critical theory. For instance, the recent analyses by Truettner, Wallach, and Sweeney in *Thomas Cole* (which I discuss in the postscript) require a more critical exposure of the ideological, political, and economic dimensions.

18. On the religious significance of the trees and mountains, see J. Gray Sweeney, "The Nude of Landscape Painting: Emblematic Personification in the Art of the Hudson River School," in *Smithsonian Studies in American Art* 3, no. 4 (Fall 1989): 43–65.

19. Letter of Cole, 1846, quoted in Louis Legrand Noble, *The Life and Works of Thomas Cole,* ed. Elliot S. Versell (Cambridge, Mass.: Harvard University Press, 1964), 82.

20. *Essay on American Scenery,* in McCoubrey, *American Art,* 98–109.

21. See the new work on this topic by J. Gray Sweeney, "'Endowed with Rare Genius': Frederic Edwin Church's *To the Memory of Cole,*" *Smithsonian Studies in American Art* 2, no. 1 (Winter 1988): 45–72, and "The Advantages of Genius," especially 114–118.

22. On Church's "inheritance" from Cole and the use he makes of the landscape and symbolism, and on the influence of Cole's work and rep-

utation during the decade after his death, see Wallach's "Thomas Cole," and Sweeney's "The Advantages of Genius."

23. *Twilight in the Wilderness* also needs to be interpreted in terms of apocalypse and the Civil War. See, for example, David Huntington's observation that the work shows "the supreme moment in cosmic time. [It] was the natural apocalypse" (*The Landscapes of Frederic Edwin Church: Vision of an American Era* [New York: Braziller, 1966], 82).

24. Fitz Hugh Ludlow, "Seven Weeks in the Great Yo-semite," *Atlantic Monthly,* June 1864, 740.

25. W. H. Holmes, "The Mountain of the Holy Cross," *Illustrated Christian Weekly,* May 1, 1875, 209.

26. Clarence King, *Mountaineering in the Sierra Nevadas* (New York: Scribner's, 1902 [1872]), 223.

27. Ibid., 364–365. Compare 114, 156, 173, 220–221, 227 ff., 237, 293, and 363.

28. See the familiar studies by Roderick Nash, *Wilderness and American Mind* (New Haven, Conn.: Yale University Press, 1967), now available in a new, expanded edition; Marjorie Hope Nicolson, *Mountain Gloom and Mountain Glory: The Development of the Aesthetics of the Infinite* (New York: Norton, 1959).

29. Although I found it too late to make use of it, Jeremy Cohen's interesting analysis of the course of Gen. 1:28 from antiquity through the Reformation is a welcome and valuable contribution to the background of this topic: *"Be Fertile and Increase, Fill the Earth and Master It": The Ancient and Medieval Career of a Biblical Text* (Ithaca, N.Y.: Cornell University Press, 1989).

30. Eliade, "Paradise and Utopia," 94; Sanford, *Quest for Paradise,* 87; Williams, *Wilderness and Paradise,* 108. The apparent mandate in Gen. 1:28 for overcoming and controlling nature may be clearer to the contemporary reader in the more recent translation of the Jerusalem Bible: "God blessed them, saying to them, 'Be fruitful, multiply, fill the earth and conquer it. Be masters of the fish of the sea, the birds of heaven and all living animals on the earth.'"

31. Eliade, "Paradise and Utopia"; Arthur A. Ekirch, *The Idea of Progress in America, 1815–1860* (New York: Columbia University Press, 1944); Franco Ferrarotti, *The Myth of Inevitable Progress* (Westport, Conn.: Greenwood, 1985).

32. Henry T. Tuckerman, *Book of the Artists: American Artist Life Comprising Biographical and Critical Sketches of American Artists* (New York: James E. Carr, 1967 [1867]), 421.

33. William Gilpin, *The Mission of the North American People, Geographical, Social, and Political,* 2d ed. (Philadelphia: Lippincott, 1873), 124.

34. Ibid., 28.

35. Ibid., 8.

36. See Leo Marx, *The Machine in the Garden: Technology and the Pastoral Ideal in America* (New York: Oxford University Press, 1964); Lewis Mumford, *Technics and Civilization* (New York: Harcourt, Brace, 1932); and Perry Miller, *Errand into the Wilderness* (Cambridge, Mass.: Harvard University Press, 1956).

37. Quoted in Marx, *Machine in the Garden,* 181.

38. Eliade, "Paradise and Utopia," 99.

39. Schwartz and Ehrenberg, *Mapping of America,* 138, 144.

40. Kynaston McShine, ed., *The Natural Paradise: Painting in America, 1800–1950* (New York: Museum of Modern Art, 1976), 87.

41. For example, consider the controversy generated by Lynn White's essay "The Historical Roots of Our Ecological Crisis" and the responses of John Macquarrie, James Barr, and others. White's essay is reprinted in *Ecology and Religion in History,* ed. David Spring and Ellen Spring (New York: Harper and Row, 1974), 15–31, as are many of the responses.

42. The hermeneutically disclosed grounds for understanding American landscape and parks as more indigenous than often is realized complements the similar argument that Galen Cranz makes on quite different grounds and with other interests. She focuses on the antiurban character of parks and on the American contribution to the picturesque; see Cranz, *The Politics of Park Design: A History of Urban Parks in America* (Cambridge, Mass.: MIT Press, 1982), 3 and 260n38; Dieter Hennebo, *Geschichte der deutschen Gartenkunst* (Hamburg: Alfred Hoffman, Broschek Verlag, 1963).

43. See, for example, Albert Fein, "The American City: The Ideal and the Real," in *The Rise of an American Architecture,* ed. Edgar Kaufmann (New York: Praeger, 1970), 51–114; Fein, *Frederick Law Olmsted and the American Environmental Tradition* (New York: Braziller, 1972); and Fein, *Landscape into Cityscape: Frederick Law Olmsted's Plans for a Greater*

New York City (Ithaca, N.Y.: Cornell University Press, 1967). Fein is an exception in that he raises the issue (although he does not pursue the background to understand more deeply the nineteenth-century cultural phenomena). The more typical approach to which I am referring is found in S. B. Sutton, *Civilizing American Cities* (Cambridge, Mass.: MIT Press, 1971). Sutton dismisses even Olmsted's "talk of social, moral, and physical benefits of parkland" in favor of factors of formal implications for design (17). How much more forgotten, then, is the dynamic *behind* this rhetoric and congruent planning principles; with Sutton we have good scholarship in presenting Olmsted's writings but a case of displacement from the meaning of our tradition just when interested designers and planners are eager to learn about it.

44. See Lee Clark Mitchell, *Witnesses to a Vanishing America: The Nineteenth-Century Response* (Princeton, N.J.: Princeton University Press, 1981), chap. 2, especially 50 ff.; Roderick Nash, "The American Invention of National Parks," *American Quarterly* 22 (Fall 1970): 726–735; Paul Herman Buck, *The Evolution of the National Park System of the United States* (Washington, D.C.: GPO, 1946); Thurman Wilkins, *Thomas Moran: Artist of the Mountains* (Norman: University of Oklahoma Press, 1966).

45. Fein, *Olmsted and the American Environmental Tradition,* 42.

46. Paul Shepard, *Man in the Landscape: A Historic View of the Esthetics of Nature* (New York: Alfred Knopf, 1967), 179; on Niagara Falls specifically, also see 141–146, 149–150, 175–176. On the general European and American spiritual response to wilderness, see Shepard's chapter 5, "The Virgin Dream," especially 188.

47. Franklin Kelly, et al., *Frederic Edwin Church* (Washington, D.C.: National Gallery of Art, 1989), 50.

48. A letter of Olmsted to William Dorsheimer, May 30, 1886, quoted in Fein, *Olmsted and the American Environmental Tradition,* plate 31; compare Olmsted on Yosemite: "The first point to be kept in mind then is the preservation and maintenance as exactly as possible of the natural scenery" ("The Yellowstone Valley and the Maripose Big Trees: A Preliminary Report," in *Landscape Architecture* 43 [October 1952 (1865)]: 22). On Niagara understood biblically as the symbol of God's covenant and, consequently, nature's beneficence, see Huntington, *The Landscapes of Frederic Edwin Church,* 71; Sweeney, *Themes in American Painting,* 51.

49. Olmsted's family governess wrote about her spiritual experience in nature (the park): "Mr. Olmsted took me to the Sequoias this afternoon. The road lies up a steep ascent covered with beautiful pines and firs and after a ride of five miles through this woodland we suddenly came upon the majestic trunk of a Sequoia. The great beauty of these forest kings is as striking as their size. The bark is a rich golden brown, and immensely thick. . . . It is formed into regular carvings like the Gothic ornaments of a cathedral yet no artificial architecture ever impressed me as much as the grand and simple outlines of these wonderful creations" ("American War Letters," July 22, 1864, quoted in Fein, *Olmsted and the American Environmental Tradition,* 39).

50. Frederick Law Olmsted, "Report upon a Projected Improvement of the Estate of the College of California, at Berkeley, near Oakland" (San Francisco: Towne and Bacon, 1866), quoted in Sutton, *Civilizing American Cities,* 270–271.

51. Sutton, *Civilizing American Cities,* 273.

52. Frederick Law Olmsted, *Mount Royal* (New York: Putnam, 1881), quoted in Sutton, *Civilizing American Cities,* 212 (my emphasis).

53. Sutton, *Civilizing American Cities,* 2. For biographical information, see Charles Capen McLaughlin and Charles E. Beveridge, eds., *The Formative Years: 1822–1852,* vol. 1 of *The Papers of Frederick Law Olmsted,* 3 vols., ed. Charles Capen McLaughlin (Baltimore: Johns Hopkins University Press, 1977); Irving D. Fisher, *Frederick Law Olmsted and the City Planning Movement in the United States* (Ann Arbor, Mich.: UMI Research Press, 1986); Albert Fein, "Frederick Law Olmsted: His Development as a Theorist and Designer of the American City" (Ph.D. Diss., Columbia University, 1969).

54. McLaughlin and Beveridge, *The Formative Years,* 235; also see the letters from this period in the same volume, for example, concerning teaching Sunday school, 242–244.

55. Ibid., 216–217.

56. Olmsted, letter to Brace, July 30, 1846, in McLaughlin and Beveridge, *The Formative Years,* 263; also see the letters discussing Channing and Bushnell, ibid., 230–232 and 240–242. On Brace introducing Olmsted to Bushnell's serious writing, see 68. On family relations with Bushnell, see 226. The famous image of the "Five Friends" made in New Haven in 1846 includes Frederick Law Olmsted, Charles Loring

Brace, John Hull Olmsted, Charles Trask (who also studied for the ministry after graduating from Yale), and Frederick Kingsbury.

57. Horace Bushnell, "City Plans," in *Work and Plan; or Literary Varieties* (New York, 1864), 308–336; Bushnell, *Nature and the Supernatural, as together Constituting the One System of God* (New York: Scribner's, 1858). Bushnell's other major theological works include: *Building Eras in Religion* (New York: Scribner's, 1881); *Christ in Theology* (Hartford, Conn.: Brown and Parsons, 1851); *God in Christ* (Hartford, Conn., Brown and Parsons, 1849). On Brace and the Children's Aid Society as a response to Bushnell's ideas, see James L. Machor, *Pastoral Cities: Urban Ideals and the Symbolic Landscape of America* (Madison: University of Wisconsin, 1987), 254.

58. Bushnell, *God in Christ,* 30–33. On Bushnell's theology, biblical interpretation, and influence, see Philip Gura, *The Wisdom of Words,* especially 51–71; H. Shelton Smith, ed., *Horace Bushnell* (New York: Oxford University Press, 1965); Barbara Cross, *Horace Bushnell: Minister to a Changing America* (Chicago: University of Chicago Press, 1958).

59. Philip Gura, *The Wisdom of Words,* 67.

60. Bushnell, "City Plans," 333, 336; quoted in Machor, *Pastoral Cities,* 146–147.

61. McLaughlin and Beveridge, *The Formative Years,* 8. See also 226–227 and the corresponding letters.

62. Ibid., 336–337, and the corresponding letters.

63. Fein, *Olmsted and the American Environmental Tradition,* 8–9; cf. 19.

64. See, for example, the argument in Bercovitch, *The American Jeremiad.*

65. Fein, *Olmsted and the American Environmental Tradition,* 55.

66. Parke Godwin, "Future of the Republic," manuscript, 51, Bryan-Godwin Papers, quoted in Fein, *Olmsted and the American Environmental Tradition,* 19. Also see Charles Loring Brace, *Home-Life in Germany* (New York, 1856), 251; Horace Bushnell, *The Principles of National Greatness* (New Haven, Conn., 1837), 14; and Bushnell, "City Plans," 308–336.

67. Frederick Law Olmsted, "Public Parks and the Enlargement of Towns," American Social Science Association (Cambridge, Mass.: Riverside, 1870), quoted in Sutton, *Civilizing American Cities,* 75.

68. Sutton, *Civilizing American Cities,* 96.

69. Hence, as Galen Cranz points out, religious services and the differences they manifested generally were forbidden in parks in the name of moral-civic homogeneity; see *The Politics of Park Design,* 23.

70. This raises the issue of the relation of urban parks to national parks, specifically in regard to whether the latter are subsumed under civic vision or remain apart, although less fundamentally important.

71. Olmsted approvingly quotes a woman who decries the benefits of country living: "If I were offered a deed of the best farm that I ever saw, on condition of going back to the country to live, I would not take it. I would rather face starvation in town" (quoted in "Public Parks," in Sutton, *Civilizing American Cities,* 58).

72. Quoted in Fein, *Olmsted and the American Environmental Tradition,* 54.

73. On the relation of Olmsted to Asa Gray, Herbert Spencer, and Lester F. Warren, see Fein, *Olmsted and the American Environmental Tradition,* 47 ff. and 53–55.

74. Cranz's *Politics of Park Design* is a major source of insight on these issues in relation to the ideologies of pleasure and reform-control.

75. Cranz, *Politics of Park Design,* vii.

76. For example, the issue is important in the current debate concerning the relation of Christianity to the exploitation of the New World and, as is increasingly realized, to what is a much more complex heritage that may also involve the roots for a spiritual ecology. See Spring and Spring, *Ecology and Religion in History*; John Carmody, *Ecology and Religion: Toward a New Christian Theology of Nature* (New York: Paulist, 1983); John Hart, *The Spirit of the Earth: A Theology of the Land* (New York: Paulist, 1984); Belden C. Lane, *Landscapes of the Sacred: Geography and Narrative in American Spirituality* (New York: Paulist, 1988); Matthew Fox, *Creation Spirituality* (New York: Harper and Row, 1989). See also Anne Primavesi, *From Apocalypse to Genesis: Ecology, Feminism, and Christianity* (Tunbridge Wells, England: Burns and Oats, 1991); Ian Bradly, *God Is Green* (London: Darton, Longman, and Todd, 1990); Lawrence E. Johnson, *A Morally Deep World: An Essay on Moral Significance and Environmental Ethics* (Cambridge: Cambridge University Press, 1991).

77. See Martin Heidegger, "The Onto-theo-logical Constitution of Metaphysics," in *Identity and Difference,* trans. Joan Stambaugh (New York: Harper and Row, 1969), 42–74, and *The Piety of Thinking: Essays*

by Martin Heidegger, trans. James G. Hart and John C. Maraldo (Bloomington: Indiana University Press, 1976).

78. In traditional analyses, since God's essence and existence, in principle, cannot differ, he is understood as the uniquely "self-sustaining act of being." See, for example, James F. Anderson, *Natural Theology: The Metaphysics of God* (Milwaukee: Bruce, 1961).

79. There is a huge and growing literature on the topic, some of it derived from reinterpretations such as Heidegger's and some from more traditional historical approaches. See, for instance, Dale Van Every, *The Disinherited: The Lost Birthright of the American Indian* (New York: Avalon, 1966); Robert F. Berkhofer, Jr., *The White Man's Indian: Images of the American Indian from Columbus to the Present* (New York: Knopf, 1978); Richard Drinnon, *Facing West: The Metaphysics of Indian Hating and Empire Building* (Minneapolis: University of Minnesota Press, 1980); Jamake Highwater, *The Primal Mind: Vision and Reality in Indian America* (New York: New American Library, 1981); J. Donald Hughs, *American Indian Ecology* (El Paso: Texas Western Press, 1983); Calvin Martin, ed., *The American Indian and the Problem of History* (New York: Oxford University Press, 1987).

80. An especially crucial ontological issue derived from the metaphysics of humans as made in God's image and likeness (as specified in Genesis), with all other creatures except angels placed lower than humans on the great chain of being. Thus, the debates as to what rights and powers extended to non-Europeans and about whether Native Americans (and blacks) were human or nonhuman (and thus subhuman) were foundationally instances of the question about metaphysical classification.

The "problem of recognition" and the need for a system of classification that resulted when attempts to use European language to describe the New World foundered are hermeneutically decoded by Anthony Pagdem in his nonphilosophical work *The Fall of Natural Man: The American Indian and the Origins of Comparative Ethnology* (New York: Cambridge University Press, 1982). Pagdem treats the context of natural law and the Aristotelian concept of the "natural slave" as they affected European intellectuals, theologians, jurists, politicians, and missionaries.

81. On the meaning of the term *landscape* see J. B. Jackson, "The Word Itself," in *Discovering the Vernacular Landscape* (New Haven,

Conn.: Yale University Press, 1984), 1–55, and "The Vernacular Landscape," in *Landscape Meanings and Values,* ed. Edmund C. Penning-Rowsell and David Lowenthal (London: Allen and Unwin, 1986), 65–77; Edward Relph, *Rational Landscapes and Humanistic Geography* (London: Croom Helm, 1981); Paul Shepard, *Man in the Landscape;* Mugerauer, "Language and the Emergence of the Environment," in *Dwelling, Place and Environment,* ed. Seamon and Mugerauer; Cosgrove, *Social Formation and Symbolic Landscape.*

For further details of the intellectual history of "landscape" and "nature," especially in relation to modern philosophy and science, see Susan Bordo, *The Flight to Objectivity: Essays on Cartesianism and Culture* (Albany, N.Y.: SUNY Press, 1987); Hal Foster, ed., *Vision and Visuality* (Seattle: Bay, 1988); Carolyn Merchant, *The Death of Nature: Women, Ecology, and the Scientific Revolution* (New York: Harper and Row, 1980); D. G. Charlton, *New Images of the Natural in France: A Study in European Cultural History 1750–1800* (New York: Cambridge University Press, 1984); Keith Thomas, *Man and the Natural World: A History of the Modern Sensibility* (New York: Pantheon, 1983).

82. Gen. 1:12, 18, 21, 25, and 31. See also Gerhard von Rad, *Genesis: A Commentary* (Philadelphia: Westminster, 1956), 44–59.

83. Heidegger's ground-breaking interpretation of *physis* is first worked out at length in *An Introduction to Metaphysics,* trans. Ralph Manheim (New York: Doubleday, 1961), chapter 4. Good commentaries are J. L. Metha, *The Philosophy of Martin Heidegger* (New York: Harper and Row, 1971), and George Seidel, *Martin Heidegger and the Pre-Socratics: An Introduction to His Thought* (Lincoln: University of Nebraska Press, 1964).

84. Luther Standing Bear, *Land of the Spotted Eagle* (Lincoln: University of Nebraska Press, 1978), 38.

85. Black Elk, *The Sacred Pipe,* ed. Joseph Epes Brown (New York: Penguin, 1973), 13–14. See also Hughs, *American Indian Ecology,* 18–19.

86. On the meaning of *wakan,* see Chunksa Yuha and James E. Ricketson, "Glossary of Lakotah Words," in Ruth Beebe Hill, *Hanta Yo* (New York: Warner, 1979), 1093–1109.

87. Ibid.

88. On the Navaho word, see Gladys A. Reichard, *Navaho Religion: A Study of Symbolism* (Princeton, N.J.: Princeton University Press,

1950), especially 507–508, where she cites her unpublished field investigations of the "male shooting chant holy and evil." Fr. Berard Haile, *Origin Legend of the Navaho Flintway* (Chicago: University of Chicago Publications in Anthropology, 1943), 13, 162.

89. On the listed words, see Jamake Highwater, *The Primal Mind*, 82–83; Hughs, *American Indian Ecology*, 18–19; Yuha and Ricketson, "Glossary of Lakotah Words."

90. On the anthropological debacle concerning *mana*, proceeding from the debate of Marrett, Tylor, and others, see Eliade, *Patterns in Comparative Religion*, where he also compares *mana* to *wakan, orenda, oki*, and the West Indian *zemi* and Bambuti (African pygmy) *megbe*.

91. John D. Caputo makes an interesting parallel point in using Derrida to broaden Heidegger's rejection of humanism as the measure by arguing that releasement "ought to be openness to all life, not just human life. . . . Letting life be *[Gelassenheit]* extends across the spectrum of living things in a generalized *Gelassenheit* (*Radical Hermeneutics*, 309).

92. The analysis of this final concealment has benefited from John D. Caputo's treatment of the disclosure and concealment of being that happened for and with Aquinas; see his *Heidegger and Aquinas: An Essay on Overcoming Metaphysics* (New York: Fordham University Press, 1982) and *Radical Hermeneutics*.

93. Heidegger's understanding of *Ereignis* (the epochal-historical development of the world) in regard to the unfolding essence of the technological era and environmental interpretation is worked out at length in my *Interpretations on Behalf of Place*, chaps. 6 through 8, 93–150.

94. Gerald L. Bruns gives an especially thoughtful reminder of the importance of concealment and scattering, which tend to be ignored even by Heidegger's readers, in his *Heidegger's Estrangements: Language, Truth and Poetry in the Later Writings* (New Haven, Conn.: Yale University Press, 1989).

95. On deep ecology see Bill Devall and George Sessions, *Deep Ecology: Living as if Nature Mattered* (Salt Lake City, Nev.: Peregine Smith, 1985), and the philosophically sophisticated work of David Rothenberg, such as *Is It Painful to Think: Conversations with Arne Naess* (Minneapolis: University of Minnesota Press, 1990) and *Hand's End: Technology and the Limits of Nature* (Berkeley: University of California

Press, 1993). An interesting philosophical hermeneutics appears in Max Oelschlager's book *The Idea of Wilderness: From Prehistory to the Age of Ecology* (New Haven, Conn.: Yale University Press, 1991). Among the many works emerging concerning the relation of Heidegger and ecology, see Michael Haar, *The Song of the Earth: Heidegger and the Grounds of the History of Being,* trans. Reginald Lilly (Bloomington: University of Indiana Press, 1993); Ladelle McWhorter, ed., *Heidegger and the Earth: Essays in Environmental Philosophy* (Kirksville, Mo.: Thomas Jefferson University Press, 1992); and Bruce V. Foltz, *Inhabiting the Earth: Heidegger, Environmental Ethics, and the Metaphysics of Nature* (Atlantic Highlands, N.J.: Humanities Press, 1995).

96. Heidegger takes up the topic of releasement toward things [*Gelassenheit*] and openness to the mystery at the same time in "Memorial Address," in *Discourse on Thinking,* trans. John M. Anderson and E. Hans Freund (New York: Harper and Row, 1966), 54–57. The issue is developed in my *Interpretations on Behalf of Place,* part 2.

Postscript

1. For the discussion of Wittgenstein and Jung as presenting the two major options that we have today, see chap. 1, the first and last sections.

2. S. Giedion, *The Eternal Present;* Etlin, *The Architecture of Death.*

3. On the "near" see Heidegger, *Discourse on Thinking* and also "Dialogue on Language," in *On the Way to Language,* trans. Peter Hertz and Joan Stambaugh (New York: Harper and Row, 1971), 1–54.

4. William H. Truettner, ed., *The West as America* (Washington, D.C.: Smithsonian Institution Press, 1991).

5. Truettner and Wallach, *Thomas Cole.*

6. Ibid., 137–138.

7. Ibid., 155.

8. Ibid., 8.

9. Wallach, "Thomas Cole," 23–112.

10. Ibid., 51.

11. Sweeney, "The Advantages of Genius," 113–135.

12. Sweeney, *The Columbus of the Woods.*

13. Cosgrove, *Social Formation and Symbolic Landscape;* Cosgrove and Daniels, *The Iconography of Landscape;* for this description see the abstract to "The Geometry of Landscape: Practical and Speculative Arts

in Sixteenth-Century Venetian Land Territories," in *The Iconography of Landscape,* 254–276.

14. Edward W. Soja, *Postmodern Geographies;* Michael Dear, "Taking Los Angeles Seriously: Time and Space in the Postmodern City," *Architecture California,* August 1991, 36–42; Manuel Castells, *The Informational City: Information Technology, Economic Restructuring, and the Urban-Regional Process* (Cambridge, England: Blackwell, 1989).

15. Jean Baudrillard, *America* and *Cool Memories,* trans. Chris Turner (New York: Verso, 1990); Jean-François Lyotard, "The Sublime and the Avant-Garde," "Scapeland," and "Newman: The Instant," in *The Lyotard Reader,* ed. Andrew Benjamin, (Oxford: Blackwell, 1989), 196–211, 212–219, 240–249, and *The Postmodern Condition.* See Mugerauer, "The Post-Structuralist Sublime: From Heterotopia to Dwelling?" a videotaped lecture presented to the Department of Landscape Architecture at the University of Minnesota, January 1991, and at the University of Washington, February 1993.

16. Ann Bermingham, *Landscape and Ideology: The English Rustic Tradition 1740–1860* (Berkeley: University of California Press, 1986); Patricia Limerick, *The Legacy of Conquest: The Unbroken Past of the American West* (New York: Norton, 1987); Alexander Wilson, *The Culture of Nature: North American Landscape from Disney to the Exxon Valdez* (Cambridge, England: Blackwell, 1992); John D. Dorst, *The Written Suburb: An American Site, An Ethnographic Dilemma* (Philadelphia: University of Pennsylvania Press, 1989); Sharon Zukin, *Landscapes of Power: From Detroit to Disney World* (Berkeley: University of California Press, 1991); Randall H. McGuire and Robert Paynter, eds., *The Archaeology of Inequality* (Oxford: Blackwell, 1991). Stephen Greenblatt, in *Marvelous Possessions: The Wonder of the New World* (Chicago: University of Chicago Press, 1991), inventively examines how Europeans represented and appropriated the lands of other peoples through colonial manipulation of representational discourse and interpretive acts and concepts; W. J. T. Mitchell, ed., *Landscape and Power* (Chicago: University of Chicago Press, 1994).

17. Michael P. Conzen, ed., *The Making of the American Landscape* (Boston: Unwin Hyman, 1990).

18. David Wyatt, *The Fall into Eden: Landscape and Imagination in California* (Cambridge: Cambridge University Press, 1986); Anne Hyde, *An American Vision: Far Western Landscapes and National Culture*

1820–1920 (New York: New York University Press, 1990). The use of paintings and the visual arts by, for example, cultural geographers displays a range of familiarity and success with the complex and historical attitudes toward nature and the works of art. See, for example, the interesting but not complete methods and views of David Lowenthal, "English Landscape Tastes," *Geographical Review* 55 (1965): 186–222, and "American Scene," in *Geographic Perspectives on America's Past,* ed. David Ward (New York: Oxford University Press, 1979), 17–32; Denis E. Cosgrove, "John Ruskin and the Geographical Imagination," *Geographical Review* 69, no. 1 (January 1972): 43–62; Ronald Rees, "Landscape in Art," in *Dimensions of Human Geography,* ed. Karl W. Butzer, et al., University of Chicago Department of Geography Research Paper 186 (Chicago: University of Chicago, 1978), 48–68; James Vance, "California and the Search for the Ideal," *Annals of the Association of American Geographers* 62 (Jan. 1972): 185–210. Especially fruitful are Kevin Starr's *Americans and the California Dream 1850–1915* (Santa Barbara, Cal.: Peregrine Smith, 1973) and *Inventing the Dream: California through the Progressive Era* (New York: Oxford University Press, 1985).

19. See Heidegger's comments on scholarship throughout *What Is Called Thinking?* trans. J. Glenn Gray and Fred D. Wieck (New York: Harper and Row, 1968).

20. Michel Foucault works on discontinuities throughout his corpus, for example, in *History of Sexuality, vol. 1: An Introduction,* trans. Robert Hurley (New York: Pantheon, 1978); on the explicit point see Dews, *Logics of Disintegration,* 109.

21. John Wilmerding, ed., *American Light: The Luminist Movement* (Washington, D.C.: National Gallery of Art, 1986); Novak, *Nature and Culture.*

22. Rees, "Landscape in Art," 186. Of course, at a metalevel, the very ideas of place and value-neutral space are constructed within the context of cultural interpretations, such as those of Cartesian and Newtonian science and epistemology. See, for instance, Max Jammer, *Concepts of Space: The History of Theories of Space in Physics* (Cambridge, Mass.: Harvard University Press, 1969).

Bibliography

Ackerman, Robert J. *Wittgenstein's City.* Amherst: University of Massachusetts Press, 1988.

Anderson, James F. *Natural Theology: The Metaphysics of God.* Milwaukee, Wis.: Bruce, 1961.

Architect's International 179, no. 7 (Feb. 15, 1984): 38.

Attoe, Wayne. *Architecture in Hot and Humid Climates.* Forthcoming.

Barr, James. "Man and Nature: The Ecological Controversy and the Old Testament." In *Ecology and Religion in History,* edited by David Spring and Ellen Spring, 48–75. New York: Harper and Row, 1974.

Baudrillard, Jean. *America.* Translated by Chris Turner. New York: Verso, 1988.

———. *Cool Memories.* Translated by Chris Turner. New York: Verso, 1990.

———. *The Mirror of Production.* Translated by Mark Poster. St. Louis: Telos, 1975.

———. *Simulations.* Translated by Paul Foss, Paul Patton, and Philip Beitchman. New York: Semiotext(e), 1983.

Beebe Hill, Ruth. *Hanta Yo.* New York: Warner, 1979.

Benedikt, Michael. *Deconstructing the Kimbell.* New York: SITES/Lumen, 1991.

Benjamin, Walter. *Das Passagen Werk* (The Arcades project; 1982). Edited by Rolf Tiedemann. Vol. 5 of *Gesammelte Schriften*. Edited by Rolf Tiedemann and Hermann Schweppenhäuser. 6 vols. Frankfurt am Main: Suhrkamp, 1972.

———. *Reflections: Essays, Aphorisms, Autobiographical Writings*. Edited by Peter Demetz and translated by Edmund Jephcott. New York: Schocken, 1978.

Bercovitch, Sacvan. *The American Jeremiad*. Madison: University of Wisconsin Press, 1978.

———. *Typology and Early American Literature*. Boston: University of Massachusetts Press, 1971.

Berkhofer, Robert F., Jr. *The White Man's Indian: Images of the American Indian from Columbus to the Present*. New York: Knopf, 1978.

Bermingham, Ann. *Landscape and Ideology: The English Rustic Tradition 1740–1860*. Berkeley: University of California Press, 1986.

Beveridge, Charles E. "Frederick Law Olmsted: The Formative Years 1822–1865." Ph.D. diss., University of Wisconsin, 1966.

Binkley, Timothy. *Wittgenstein's Language*. Ph.D. diss., University of Texas at Austin, 1970.

———. *Wittgenstein's Language*. The Hague: Nijhoff, 1973.

Black Elk. *The Sacred Pipe*. Edited by Joseph Epes Brown. New York: Penguin, 1973.

Blondel, Jacques-François. *L'Architecture française (1752–56)*. Edited by Louis Savot. Reprint. Geneva: Minkoff, 1973.

Bordo, Susan. *The Flight to Objectivity: Essays on Cartesianism and Culture*. Albany: SUNY Press, 1987.

Bouwsma, O. K. *Philosophical Essays*. Lincoln: University of Nebraska Press, 1965.

Boyer, M. Christine. *Dreaming the Rational City: The Myth of American City Planning*. London: MIT Press, 1983.

Brace, Charles Loring. *Home-Life in Germany*. New York, 1856.

Bradley, Ian. *God Is Green*. London: Darton, Longman, and Todd, 1990.

Bridges, Bonnie. "Feminism, Civic Architecture, and Political Philosophy." Master's thesis, University of Texas at Austin, 1990.

Bridges, Bonnie, and Robert Mugerauer. "Recasting the Body Politic: Transformations of the Athenian Agora." In *Bodies: Image, Writing, Technology,* edited by Juliet F. MacCannell and Laura Zakarin. Albany, N.Y.: SUNY Press, forthcoming.

Brown, Jerry Wayne. *The Rise of Biblical Criticism in America.* Middletown, Conn.: Wesleyan University Press, 1969.

Brumm, Ursula. *American Thought and Religious Typology.* Translated by John Hoaglund. New Brunswick, N.J.: Rutgers University Press, 1970.

Bruns, Gerald. *Heidegger's Estrangements: Language, Truth and Poetry in the Later Writings.* New Haven, Conn.: Yale University Press, 1989.

Buck, Paul Herman. *The Evolution of the National Park System of the United States.* Washington, D.C.: GPO, 1946.

Buck-Morss, Susan. *The Dialectics of Seeing: Walter Benjamin and the Arcades Project.* Cambridge, Mass.: MIT Press, 1989.

Burckhardt, Titus. *Sacred Art in East and West.* Translated by Lord Northbourne. Bedford, England: Perennial, 1967.

Bushnell, Horace. *Building Eras in Religion.* New York: Scribner's, 1881.

———. *Christ in Theology.* Hartford, Conn.: Brown and Parsons, 1851.

———. "City Plans." In *Work and Plan; or Literary Varieties,* 308–336. New York, 1864.

———. *God in Christ.* Hartford, Conn.: Brown and Parsons, 1849.

———. *Nature and the Supernatural, as together Constituting the One System of God.* New York: Scribner's, 1858.

———. *The Principles of National Greatness.* New Haven, Conn., 1837.

Buttimer, Anne. "Home, Reach, and Sense of Place." In *The Human Experience of Space and Place,* edited by A. Buttimer and D. Seamon, 166–187. London: Croom Helm, 1980.

Buttimer, Anne, and David Seamon, eds. *The Human Experience of Space and Place.* London: Croom Helm, 1980.

Caputo, John D. *Heidegger and Aquinas: An Essay on Overcoming Metaphysics.* New York: Fordham University Press, 1982.

———. *Radical Hermeneutics: Repetition, Deconstruction, and the Hermeneutic Project.* Bloomington: Indiana University Press, 1987.

Carmody, John. *Ecology and Religion: Toward a New Christian Theology of Nature.* New York: Paulist, 1983.

Castells, Manuel. *The Informational City: Information Technology, Economic Restructuring, and the Urban-Regional Process.* Cambridge, Mass.: Blackwell, 1989.

———. *The Space Question.* Oxford, England: Blackwell, 1990.

Charlton, D. G. *New Images of the Natural in France: A Study in European Cultural History 1750–1800.* New York: Cambridge University Press, 1984.

Clebsch, William. *From Sacred to Profane America: The Role of Religion in American History.* New York: Harper and Row, 1968.

Cohen, Jeremy. *"Be Fertile and Increase, Fill the Earth and Master It": The Ancient and Medieval Career of a Biblical Text.* Ithaca, N.Y.: Cornell University Press, 1989.

Cole, Thomas. *Essay on American Scenery.* In *American Art, 1700–1960,* edited by John W. McCoubrey, 98–109. Englewood Cliffs, N.J.: Prentice-Hall, 1965.

Conzen, Michael P., ed. *The Making of the American Landscape.* Boston: Unwin Hyman, 1990.

Cooper, Clare. "The House as Symbol of Self." In *Designing for Human Behavior: Architecture and the Behavioral Sciences,* edited by Jon Lang et al., 130–146. Stroudsburg, Pa.: Dowden, Hutchinson, & Ross, 1974.

Cosgrove, Denis E. "John Ruskin and the Geographical Imagination." *Geographical Review* 69, no. 1 (January 1972): 43–62.

———. *Social Formation and Symbolic Landscape.* Totowa, N.J.: Barnes and Noble, 1984.

Cosgrove, Denis E., and Stephen Daniels, eds. *The Iconography of Landscape: Essays on the Symbolic Representation, Design, and Use of Past Environments.* New York: Cambridge University Press, 1988.

Cranz, Galen. *The Politics of Park Design: A History of Urban Parks in America.* Cambridge, Mass.: MIT Press, 1982.

Cronon, William. *Changes in the Land: Indians, Colonists, and the Ecology of New England.* New York: Hill and Wang, 1983.

Cross, Barbara. *Horace Bushnell: Minister to a Changing America.* Chicago: University of Chicago Press, 1958.

Crow, Dennis, ed. *Philosophical Streets.* Washington, D.C.: Maisonneuve, 1990.

Davidson, James West. *The Logic of Millennial Thought: Eighteenth-Century New England*. New Haven, Conn.: Yale University Press, 1977.

Dear, Michael. "Taking Los Angeles Seriously: Time and Space in the Postmodern City." *Architecture California* (Aug. 1991): 36–42.

de Certeau, Michel. *Heterologies: Discourse on the Other*. Translated by Brian Massumi. Minneapolis: University of Minnesota Press, 1989.

———. *The Practice of Everyday Life*. Translated by Steven Randall. Berkeley: University of California Press, 1984.

Deleuze, Gilles. *The Logic of Sense*. Translated by Mark Lester. New York: Columbia University Press, 1989.

Deleuze, Gilles, and Félix Guattari. *A Thousand Plateaus: Capitalism and Schizophrenia*. Translated by Brian Massumi. Minneapolis: University of Minnesota Press, 1988.

Derr, Thomas S. "Religious Responsibility for the Ecological-Crisis: An Argument Run Amok." *Worldview* 18 (Jan. 1975): 39–45.

Derrida, Jacques. "Differance." In *Speech and Phenomena,* translated by David B. Allison, 129–160. Evanston, Ill.: Northwestern University Press, 1973.

———. "In Discussion with Christopher Norris." In *Deconstruction: The Omnibus Volume,* 71–75. New York: Rizzoli, 1989.

———. *Dissemination*. Translated by Barbara Johnson. Chicago: University of Chicago Press, 1981.

———. *Edmund Husserl's Origin of Geometry: An Introduction*. Translated by John P. Leavey, Jr. Stony Brook, N.Y.: Nicholas Hays, 1978.

———. *Of Grammatology*. Translated by Gayatri Spivak. Baltimore: Johns Hopkins University Press, 1976.

———. Interview. *Domus* 671 (April 1986): 17–24.

———. "A Letter to Peter Eisenman." *Assemblage* 12 (1990): 7–13.

———. *Margins of Philosophy*. Translated by Alan Bass. Chicago: University of Chicago Press, 1982.

———. *Memoirs of the Blind: The Self-Portrait and Other Ruins.* Translated by Pascale-Anne Brault and Michael Nass. Chicago: University of Chicago Press, 1993.

———. "The Pit and the Pyramid: Introduction to Hegel's Semiology." In *Margins of Philosophy,* translated by Alan Bass, 69–108. Chicago: University of Chicago Press, 1982.

———. *Positions.* Translated by Alan Bass. Chicago: University of Chicago Press, 1981.

———. "Restitutions of the Truth in Pointing [*pointure*]." In *The Truth of Painting,* translated by Geoff Bennington and Ian McLeod, 255–382. Chicago: University of Chicago Press, 1987.

———. *Speech and Phenomena.* Translated by David Allison. Evanston, Ill.: Northwestern University Press, 1973.

———. *Spurs: Nietzsche's Styles.* Translated by Barbara Harlow. Chicago: University of Chicago Press, 1978.

———. *The Truth in Painting.* Translated by Geoff Bennington and Ian McLeod. Chicago: University of Chicago Press, 1987.

———. "Why Peter Eisenman Writes Such Good Books." In *Threshold* 4 (Spring 1988): 99–105.

———. *Writing and Difference.* Translated by Alan Bass. Chicago: University of Chicago Press, 1978.

Devall, Bill, and George Sessions. *Deep Ecology: Living as if Nature Mattered.* Salt Lake City, Nev.: Peregrine Smith, 1985.

Dews, Peter. *Logics of Disintegration: Post-Structuralist Thought and the Claims of Critical Theory.* New York: Verso, 1987.

Dorst, John D. *The Written Suburb: An American Site, An Ethnographic Dilemma.* Philadelphia: University of Pennsylvania Press, 1989.

Doughty, Robin. *At Home in Texas: Early Views of the Land.* College Station: Texas A&M University Press, 1983.

———. *Wildlife and Man in Texas: Environmental Change and Conservation.* College Station: Texas A&M Press, 1987.

Drinnon, Richard. *Facing West: The Metaphysics of Indian Hating and Empire Building.* Minneapolis: University of Minnesota Press, 1980.

Eisenman, Peter. "Post/El Cards: A Reply to Jacques Derrida." *Assemblage* 12 (1990): 14–17.

Ekirch, Arthur A. *The Idea of Progress in America, 1815–1860.* New York: Columbia University Press, 1944.

Eliade, Mircea. *Cosmos and History: The Myth of the Eternal Return.* New York: Harper and Row, 1959.

————. *Myth and Reality*. Translated by Willard R. Trask. New York: Harper and Row, 1963.

————. *Myths, Dreams, and Mysteries: The Encounter between Contemporary Faiths and Archaic Realities*. Translated by Philip Mairet. New York: Harper and Row, 1960.

————. "Paradise and Utopia: Mythical Geography and Eschatology." In *The Quest* 88–111. Chicago: University of Chicago Press, 1969.

————. *Patterns in Comparative Religion*. Translated by Rosemary Sheed. New York: World, 1958.

————. *The Quest: History and Meaning in Religion*. Chicago: University of Chicago Press, 1969.

————. *The Sacred and the Profane: The Nature of Religion*. Translated by Willard R. Trask. New York: Harper and Row, 1959.

————. *The Two and the One*. Translated by J. M. Cohen. New York: Harper and Row, 1965.

Etlin, Richard A. *The Architecture of Death: The Transformation of the Cemetery in Eighteenth-Century Paris*. Cambridge, Mass.: MIT Press, 1984.

Fein, Albert. "The American City: The Ideal and the Real." In *The Rise of an American Architecture,* edited by Edgar Kaufmann, 51–114. New York: Praeger, 1970.

————. *Frederick Law Olmsted and the American Environmental Tradition*. New York: Braziller, 1972.

————. "Frederick Law Olmsted: His Development as a Theorist and Designer of the American City." Ph.D. diss., Columbia University, 1969.

————. *Landscape into Cityscape: Frederick Law Olmsted's Plans for a Greater New York City*. Ithaca, N.Y.: Cornell University Press, 1967.

Ferrarotti, Franco. *The Myth of Inevitable Progress*. Westport, Conn.: Greenwood, 1985.

Fisher, Irving D. *Frederick Law Olmsted and the City Planning Movement in the United States*. Ann Arbor, Mich.: UMI Research Press, 1986.

Foltz, Bruce V. *Inhabiting the Earth: Heidegger, Environmental Ethics, and the Metaphysics of Nature*. Atlantic Highlands, N.J.: Humanities Press, 1995.

Foster, Hal, ed. *Vision and Visuality.* Seattle: Bay Press, 1988.

Foucault, Michel. *The Archaeology of Knowledge and the Discourse on Language.* Translated by Alan Sheridan Smith. New York: Pantheon, 1972.

————. *The Birth of the Clinic: An Archaeology of Medical Perception.* Translated by Alan Sheridan Smith. New York: Pantheon, 1977.

————. *Discipline and Punish: The Birth of the Prison.* Translated by Alan Sheridan Smith. New York: Vintage, 1979.

————. *History of Sexuality, vol. 1.: An Introduction.* Translated by Robert Hurley. New York: Pantheon, 1978.

————. *Madness and Civilization: A History of Insanity in the Age of Reason.* Translated by Richard Howard. New York: Vintage, 1965.

————. "Nietzsche, Freud, Marx." In *Nietzsche,* 183–200. Paris: Cahiers de Royaumont, 1967.

————. "Nietzsche, Genealogy, History." In *Language, Counter-Memory, Practice: Selected Essays and Interviews,* edited by Donald F. Bouchard and translated by D. F. Bouchard and Sherry Simon, 139–164. New York: Cornell University Press, 1977.

————. *The Order of Things: An Archaeology of the Human Sciences.* Translated by Alan Sheridan Smith. New York: Random House, 1973.

————. *Power/Knowledge: Selected Interviews and Other Writings, 1972–1977.* Edited by Colin Gordon and translated by C. Gordon et al. New York: Pantheon, 1980.

Fox, Matthew. *Creation Spirituality.* New York: Harper and Row, 1989.

Frankfurt, H. *Kingship and the Gods.* Chicago: University of Chicago Press, 1948.

Frei, Hans W. *The Eclipse of Biblical Narrative: A Study in Eighteenth and Nineteenth Century Hermeneutics.* New Haven, Conn.: Yale University Press, 1974.

Frye, Northrop. *Anatomy of Criticism.* New York: Atheneum, 1968.

————. "Levels of Meaning in Literature." *Kenyon Review* (Spring 1950): 246–262.

Gadamer, Hans-Georg. "On the Circle of Understanding." In *Hermeneutics vs. Science? Three German Views,* edited and translated by John M. Connolly and Thomas Keutner, 68–87. Notre Dame, Ind.: University of Notre Dame Press, 1988.

————. *Philosophical Hermeneutics.* Translated and edited by David E.

Linge. Berkeley: University of California Press, 1976.

———. *Truth and Method.* Translated by Garrett Barden and John Cumming. New York: Seabury, 1975 (2d rev. ed. translated by Joel Weinsheimer and Donald G. Marshall [New York: Continuum, 1989]).

Garver, Newton, and Seung-Chong Lee. *Derrida and Wittgenstein.* Philadelphia: Temple University Press, 1994.

Giedion, Sigfried. *The Eternal Present: The Beginnings of Art.* New York: Bollingen Foundation, 1964.

Gilpin, William. *The Mission of the North American People, Geographical, Social, and Political.* 2d. ed. Philadelphia: Lippincott, 1873.

Godwin, Parke. "Future of the Republic." Manuscript. Bryan-Godwin Papers, Manuscript Division, New York Public Library.

Goetzmann, William H., and William Orr. *Karl Bodmer's America.* Lincoln: University of Nebraska Press, 1984.

Goetzmann, William H., and William N. Goetzmann. *The West of the Imagination.* New York: Norton, 1986.

Goetzmann, William H., and Joseph C. Porter. *The West as Romantic Horizon.* Lincoln: Joslyn Art Museum and University of Nebraska Press, 1981.

Gravagnuolo, Benedetto, ed. *Adolf Loos: Theory and Works.* New York: Rizzoli, 1988.

Greenblatt, Stephen. *Marvelous Possessions: The Wonder of the New World.* Chicago: University of Chicago Press, 1991.

Gregory, Derek. "Human Agency and Human Geography." *Transactions, Institute of British Geographers* 6 (1981): 1–18.

———. "A Realist Construction of the Social." *Transactions, Institute of British Geographers* 7 (1982): 254–256.

Grubar, Francis S. *William Ranney.* Washington, D.C.: The Corcoran Gallery of Art, 1962.

Gura, Philip F. *The Wisdom of Words: Language, Theology, and Literature in the New England Renaissance.* Middletown, Conn.: Wesleyan University Press, 1981.

Haar, Michael. *The Song of the Earth: Heidegger and the Grounds of the History of Being.* Translated by Reginald Lilly. Bloomington: University of Indiana Press, 1993.

Habermas, Jürgen. *Knowledge and Human Interests.* Translated by

Jeremy J. Shapiro. Boston: Beacon, 1971.

———. *The Theory of Communicative Action.* Translated by Thomas McCarthy. 2 vols. Boston: Beacon, 1984, 1987.

Haile, Fr. Berard. *Origin Legend of the Navaho Flintway.* Chicago: University of Chicago Publications in Anthropology, 1943.

Harland, Richard. *Superstructuralism: The Philosophy of Structuralism and Post-Structuralism.* New York: Methuen, 1987.

Harries, Karsten. *The Bavarian Rococo Church.* New Haven, Conn.: Yale University Press, 1983.

———. "Space, Place, and Ethos: Reflections on the Ethical Function of Architecture." *Artibus et Historiae* 9 (1984): 159–165.

———. "Thoughts on a Non-Arbitrary Architecture." *Perspecta* 20 (1983): 9–20.

Hart, John. *The Spirit of the Earth: A Theology of the Land.* New York: Paulist, 1984.

Hartman, Geoffrey H. *Saving the Text: Literature/Derrida/Philosophy.* Baltimore: Johns Hopkins University Press, 1981.

Harvey, David. *The Condition of Postmodernity: An Enquiry into the Origins of Cultural Change.* Oxford, England: Blackwell, 1989.

Heidegger, Martin. *Basic Writings.* Edited by David Krell. New York: Harper and Row, 1977.

———. *Being and Time.* Translated by John M. Anderson and E. H. Freund. New York: Harper and Row, 1962 (1927).

———. *Discourse on Thinking.* Translated by John M. Anderson and E. H. Freund. New York: Harper & Row, 1966.

———. *Early Greek Thinking.* Translated by Frank Capuzzi amd David Krell. New York: Harper and Row, 1975.

———. *Identity and Difference.* Translated by Joan Stambaugh. New York: Harper and Row, 1969.

———. *An Introduction to Metaphysics.* Translated by Ralph Manheim. New York: Doubleday, 1961.

———. *The Piety of Thinking: Essays by Martin Heidegger.* Translated by James G. Hart and John C. Maraldo. Bloomington: Indiana University Press, 1976.

———. *Poetry, Language, Thought.* Translated by Albert Hofstadter. New York: Harper & Row, 1971.

———. *The Question concerning Technology and Other Essays.* Edited and translated by William Lovitt. New York: Harper & Row, 1977.

————. "Sprache und Heimat" (Language and Homeland). In *Hebbel-Jahrbuch* 11 (1970): 17–39.

————. *On Time and Being.* Translated by Joan Stambaugh. New York: Harper and Row, 1972.

————. *On the Way to Language.* Translated by Peter D. Hertz and Joan Stambaugh. New York: Harper and Row, 1971.

————. *What Is Called Thinking?* Translated by J. Glenn Gray and Fred D. Wieck. New York: Harper and Row, 1968.

Hennebo, Dieter. *Geschichte der deutschen Gartenkunst* (History of German park design). Hamburg: Alfred Hoffman, Broschek Verlag, 1963.

Hiers, Richard H. "Ecology, Biblical Theology, and Methodology: Biblical Perspectives on the Environment." *Zygon* 19, no. 1 (March 1984): 43–59.

Highwater, Jamake. *The Primal Mind: Vision and Reality in Indian America.* New York: New American Library, 1981.

Hirsch, E. D., Jr. *Validity in Interpretation.* New Haven, Conn.: Yale University Press, 1967.

Hollier, Denis. *Against Architecture: The Writings of George Bataille.* Cambridge, Mass.: MIT Press, 1989.

Holmes, William H. "The Mountain of the Holy Cross." *Illustrated Christian Weekly,* May 1, 1875, 208–210.

Hospers, John. *Meaning and Truth in the Arts.* Chapel Hill: University of North Carolina Press, 1946.

Hughs, J. Donald. *American Indian Ecology.* El Paso: Texas Western Press, 1983.

Huntington, David C. *Art and the Excited Spirit: America in the Romantic Period.* Ann Arbor: University of Michigan Museum of Art, 1972.

————. *The Landscapes of Frederic Edwin Church: Vision of an American Era.* New York: Braziller, 1966.

Hyde, Anne F. *An American Vision: Far Western Landscape and National Culture, 1820–1920.* New York: New York University Press, 1990.

Ihde, Don. *Hermeneutic Phenomenology: The Philosophy of Paul Ricoeur.* Evanston, Ill.: Northwestern University Press, 1971.

Irigaray, Luce. *Speculum of the Other Woman.* Translated by Gillian C.
Gill. Ithaca, N.Y.: Cornell University Press, 1993.

Jackson, John Brinckerhoff. *American Space: The Centennial Years:*
1865–1876. New York: Norton, 1972.

———. *Discovering the Vernacular Landscape.* New Haven, Conn.:
Yale University Press, 1984.

———. "Ghosts at the Door." *Landscape* 1, no. 2 (Autumn 1951): 2–9.

———. *Landscapes.* Edited by Ervin H. Zube. Amherst: University of
Massachusetts Press, 1970.

———. *The Necessity for Ruins and Other Topics.* Amherst: University
of Massachusetts Press, 1980.

———. "The Vernacular Landscape." In *Landscape Meanings and*
Values, edited by Edmund C. Penning-Rowsell and David
Lowenthal, 65–77. London: Allen and Unwin, 1986.

Jacobs, Paul, and Saul Landau. *To Serve the Devil.* 2 vols. New York:
Vintage, 1971.

Jameson, Fredric. "Architecture and the Critique of Ideology." In *The*
Ideologies of Theory: Essays 1971–1986, 2 vols., 2:35–60.
Minneapolis: University of Minnesota Press, 1988.

Jammer, Max. *Concepts of Space: The History of Theories of Space in*
Physics. Cambridge, Mass.: Harvard University Press, 1969.

Johnson, Hildegard B. "Toward a National Landscape." In *The Making*
of the American Landscape, edited by Michael Conzen, 127–145.
London: Unwin Hyman, 1990.

Johnson, Lawrence E. *A Morally Deep World: An Essay on Moral*
Significance and Environmental Ethics. Cambridge: Cambridge
University Press, 1991.

Johnson, Philip, and Mark Wigley. *Deconstructivist Architecture.* New
York: Museum of Modern Art, 1988.

Jukes, Peter. *A Shout in the Street: An Excursion into the Modern City.*
Berkeley: University of California Press, 1991.

Jung, Carl G. *The Archetypes of the Collective Unconscious.* Vol. 9,
part 1 of *Collected Works,* translated by R. F. C. Hull. 2d ed. 20 vols.
Princeton, N.J.: Princeton University Press, 1967–1979.

———. *Memories, Dreams, Reflections.* Edited by Aniela Jaffé. New
York: Pantheon, 1973.

————. *Modern Man in Search of a Soul.* New York: Harcourt, Brace, Jovanovich, 1955.

————. *Symbols of Transformation.* Vol. 5, part 1 of *Collected Works,* translated by R. F. C. Hull. 2d ed. 20 vols. Princeton, N.J.: Princeton University Press, 1967–1979.

————. *Two Essays on Analytical Pschology.* Vol. 7 of *Collected Works,* translated by R. F. C. Hull. 2d ed. 20 vols. Princeton, N.J.: Princeton University Press, 1967–1979.

Kelly, Franklin, et al. *Frederic Edwin Church.* Washington, D.C.: National Gallery of Art, 1989.

Ketner, Joseph D. II, and Michael J. Tammenga. *The Beautiful, the Sublime, and the Picturesque.* St. Louis: Washington University Press, 1984.

King, Clarence. *Mountaineering in the Sierra Nevadas.* New York: Scribner's, 1902 (1872).

Kipnis, Jeffrey. *Choral Work.* London: Architectural Association, forthcoming.

Kristeva, Julia. *The Kristeva Reader.* Edited by Toril Moi. Translated by Léon Roudica, Séan Handy, et al. Oxford: Blackwell, 1987.

Lane, Belden C. *Landscapes of the Sacred: Geography and Narrative in American Spirituality.* New York: Paulist, 1988.

Lang, Jon, et al., eds. *Designing for Human Behavior: Architecture and the Behavioral Sciences.* Stroudsburg, Pa.: Dowden, Hutchinson, & Ross, 1974.

Lefebvre, Henri. *Everyday Life in the Modern World.* Translated by Sacha Rabinovitch. Harmondsworth, England: Penguin, 1971.

————. *The Production of Space.* Translated by Donald Nicholson-Smith. Oxford: Blackwell, 1991.

Leitner, Bernhard. *The Architecture of Ludwig Wittgenstein: A Documentation.* New York: New York University Press, 1976.

Lewis, R. W. B. *The American Adam: Innocence, Tragedy, and Tradition in the Nineteenth Century.* Chicago: University of Chicago Press, 1955.

Limerick, Patricia Nelson. *The Legacy of Conquest: The Unbroken Past of the American West.* New York: Norton, 1987.

Loos, Adolf. *Spoken into the Void: Collected Essays 1897–1900.*

Translated by Jane O. Newman and John H. Smith. Cambridge, Mass.: MIT Press, 1982.

Lowenthal, David. "The American Scene." In *Geographic Perspectives on America's Past,* edited by David Ward, 17–32. New York: Oxford University Press, 1979.

Lowenthal, David, and H. C. Price. "English Landscape Tastes." *Geographical Review* 55 (1965): 186–222.

Ludlow, Fitz Hugh. "Seven Weeks in the Great Yo-semite." *Atlantic Monthly,* June 1864.

Lyotard, Jean-François. *Driftworks.* Translated by Roger McKeon, et al. New York: Semiotext(e).

———. *The Lyotard Reader.* Edited by Andrew Benjamin. Multiple translators. Oxford, England: Blackwell, 1989.

———. *The Postmodern Condition: A Report on Knowledge.* Translated by Geoff Bennington and Brian Massumi. Minneapolis: University of Minnesota Press, 1984.

McClintock, Jean Gardner. "Olmsted on the Road: A View of Paradise." In *Art of the Olmsted Landscape,* edited by Bruce Kelly et al., 21–29. New York: Arts, 1981.

McCoubrey, John W., ed. *American Art, 1700–1960.* Englewood Cliffs, N.J.: Prentice-Hall, 1965.

McGuire, Randall H., and Robert Paynter, eds. *The Archaeology of Inequality.* Oxford, England: Blackwell, 1991.

Machor, James L. *Pastoral Cities: Urban Ideals and the Symbolic Landscape of America.* Madison: University of Wisconsin Press, 1987.

MacIntyre, Alasdair. *Three Rival Versions of Moral Inquiry: Encyclopedia, Genealogy, and Tradition.* Notre Dame, Ind.: University of Notre Dame Press, 1989.

———. *Whose Justice? Which Rationality?* Notre Dame, Ind.: University of Notre Dame Press, 1990.

Mackey, Louis H. "Notes toward a Definition of Philosophy." *Franciscan Studies* 33, no. 11 (1973): 262–272.

McLaughlin, Charles Capen, and Charles E. Beveridge, eds. *The Formative Years: 1822–1852.* Vol. 1 of *The Papers of Frederick Law Olmsted,* edited by Charles Capen McLaughlin. 3 vols. Baltimore: Johns Hopkins University Press, 1977.

McShine, Kynaston, ed. *The Natural Paradise: Painting in America, 1800–1950.* New York: Museum of Modern Art, 1976.

McWhorter, Ladelle, ed. *Heidegger and the Earth: Essays in Environmental Philosophy.* Kirksville, Mo.: Thomas Jefferson University Press, 1992.

Malcolm, Norman. *Ludwig Wittgenstein: A Memoir.* With a biographical sketch by G. H. von Wright. New York: Oxford University Press, 1958.

Maritain, Jacques. *Art and Scholasticism.* New York: Scribner's, 1962.

Martin, Calvin, ed. *The American Indian and the Problem of History.* New York: Oxford University Press, 1987.

Marx, Leo. *The Machine in the Garden: Technology and the Pastoral Ideal in America.* New York: Oxford University Press, 1964.

Marx, Richard. "Egyptian Architecture in Los Angeles." *Los Angeles Times,* Sunday supplement, May, 10, 1977.

Merchant, Caroline. *The Death of Nature: Women, Ecology, and the Scientific Revolution.* New York: Harper and Row, 1980.

Metha, J. L. *The Philosophy of Martin Heidegger.* New York: Harper and Row, 1971.

Miller, Perry. *Errand into the Wilderness.* Cambridge, Mass.: Harvard University Press, 1956.

Mitchell, Lee Clark. *Witnesses to a Vanishing America: The Nineteenth-Century Response.* Princeton, N.J.: Princeton University Press, 1981.

Mitchell, W. J. T. *Landscape and Power.* Chicago: University of Chicago Press, 1994.

Moore, James Collins. *The Storm and the Harvest: The Image of Nature in Mid-Nineteenth Century Landscape Paintings.* Ph.D. diss., University of Indiana, 1974.

Mugerauer, Robert. "Toward an Architectural Vocabulary: The Porch as Between." In *Dwelling, Seeing, and Designing: Toward a Phenomenological Ecology,* edited by David Seamon, 103–128. Albany, N.Y.: SUNY Press, 1992.

———. "Architecture as Properly Useful Opening." In *Ethics and Danger: Essays on Heidegger and Continental Thought,* edited by Charles Scott and R. Dallery, 215–226. Albany, N.Y.: SUNY Press, 1991.

———. "Chicago's Four-Layered Plan." *Crit* 17 (1987): 17–23.

———. "Egyptian and American Pyramid Complexes." Research Paper for the General Program of Liberal Studies, University of Notre Dame, Spring 1965.

———. *Heidegger's Language and Thinking.* Atlantic Highlands, N.J.: Humanities, 1988.

———. "The Historical Dynamic of the American Landscape." Paper presented to the Council of Educators in Landscape Architecture, University of Illinois, Urbana-Champaign, September 1985.

———. *Interpretations on Behalf of Place: Environmental Displacements and Alternative Responses.* Albany, N.Y.: SUNY Press, 1994.

———. "Language and the Emergence of Environment." In *Dwelling, Place and Environment,* edited by David Seamon and Robert Mugerauer, 51–70. Dordrecht, Holland: Nijhoff, 1985; reprint, New York: Columbia University Press, Morningstar Editions, 1989.

———. "Midwestern Yards." *Places* 2, no. 2 (1985): 31–38.

———. "Phenomenology and Vernacular Architecture." In *Encyclopedia of World Vernacular Architecture,* edited by Paul Oliver. 4 vols. London: Blackwell, forthcoming.

———. "Phenomenology and The Environmental Disciplines." University of Texas Community and Regional Planning Working Paper Series. Austin: University of Texas Graduate Program in Community and Regional Planning, 1991.

———. "Post-Structuralist Planning Theory." University of Texas Graduate Program in Community and Regional Planning Working Paper Series. Austin: University of Texas Graduate Program in Community and Regional Planning, 1991.

———. "The Post-Structuralist Sublime: From Heterotopia to Dwelling?" Videotaped lecture presented to the Department of Landscape Architecture at the University of Minnesota, January 1991, and the University of Washington, February 1992.

Mumford, Lewis. *Technics and Civilization.* New York: Harcourt, Brace, 1932.

Nash, Roderick. "The American Invention of National Parks." *American Quarterly* 22 (Fall 1970): 726–735.

———. *Wilderness and the American Mind.* New Haven, Conn.: Yale University Press, 1967.

Nedo, Michael, and Michele Ranchetti, eds. *Wittgenstein—Sein Leben in Bildern und Texten* (Wittgenstein—his life in pictures and texts). Frankfurt am Main: Suhrkamp, 1983.

Nicolson, Marjorie Hope. *Mountain Gloom and Mountain Glory: The Development of the Aesthetics of the Infinite.* New York: Norton, 1959.

Noble, Louis Legrand. *The Life and Works of Thomas Cole.* Edited by Elliot S. Versell. Cambridge, Mass.: Harvard University Press, 1964.

Norberg-Schulz, Christian. *Architecture: Meaning and Place.* New York: Rizzoli, 1988.

———. *The Concept of Dwelling: On the Way to Figurative Architecture.* New York: Rizzoli, 1985.

———. *Genius Loci: Towards a Phenomenology of Architecture.* New York: Rizzoli, 1979.

———. "Heidegger's Thinking on Architecture." *Perspecta* 20 (1983): 61–68.

Novak, Barbara. *American Light.* Washington, D.C.: National Gallery of Art, 1986.

———. *Nature and Culture: American Landscape and Painting, 1825–1875.* New York: Oxford University Press, 1980.

O'C. Drury, M. "Conversations with Wittgenstein." In *Ludwig Wittgenstein: Personal Recollections,* edited by Rush Rhees. Totowa, N.J.: Rowan and Littlefield, 1981.

Oelschlager, Max. *The Idea of Wilderness: From Prehistory to the Age of Ecology.* New Haven, Conn.: Yale University Press, 1991.

Oliver, Paul, ed. *Encyclopedia of World Vernacular Architecture.* 4 vols. London: Blackwell, forthcoming.

Olmsted, Frederick Law. "Mount Royal, Montreal." New York: Putnam, 1881.

———. "Public Parks and the Enlargement of Towns." American Social Science Association. Cambridge, Mass.: Riverside, 1870.

———. "Report upon a Projected Improvement of the Estate of the College of California, at Berkeley, near Oakland." San Francisco: Towne and Bacon, 1866.

———. "The Yellowstone Valley and the Maripose Big Trees: A Preliminary Report." *Landscape Architecture* 43 (October 1952 [1865]): 12–25.

Pagdem, Anthony. *The Fall of Natural Man: The American Indian and the Origins of Comparative Ethnology.* New York: Cambridge University Press, 1982.

Palmer, Richard E. *Hermeneutics: Interpretation Theory in Schleiermacher, Dilthey, and Gadamer.* Evanston, Ill.: Northwestern University Press, 1969.

Panofsky, Erwin. *Meaning in the Visual Arts: Papers in and on Art History.* Garden City, N.Y.: Doubleday, 1955.

Papadakis, Andreas, et al., eds. *Deconstruction, the Omnibus Volume.* New York: Rizzoli, 1989.

Pastier, John. "Isozaki's Design for MOCA." *Arts and Architecture* 2, no. 1 (1983): 31–34.

Penning-Rowsell, Edmund C., and David Lowenthal, eds. *Landscape Meanings and Values.* London: Allen and Unwin, 1986.

Pérouse de Montclos, Jean-Marie. *Etienne-Louis Boullée: De L'Architecture classique à l'architecture révolutionnaire* (Boullée: from classical architecture to revolutionary architecture). Paris: Arts et Métiers Graphiques, 1969.

Primavesi, Anne. *From Apocalypse to Genesis: Ecology, Feminism, and Christianity.* Tunbridge Wells, England: Burns and Oats, 1991.

Rapoport, Amos. *House Form and Culture.* Englewood Cliffs, N.J.: Prentice-Hall, 1969.

Rees, Ronald. "Landscape in Art." In *Dimensions of Human Geography,* edited by Karl W. Butzer, et al., 48–68. University of Chicago, Department of Geography, Research Paper 186. Chicago: University of Chicago, 1978.

Reichard, Gladys A. *Navaho Religion: A Study of Symbolism.* Princeton, N.J.: Princeton University Press, 1950.

Relph, Edward. *The Modern Urban Landscape.* Baltimore: Johns Hopkins University Press, 1987.

———. *Place and Placelessness.* London: Pion, 1976.

———. *Rational Landscapes and Humanistic Geography.* London: Croom Helm, 1981.

Rhees, Rush, ed. *Ludwig Wittgenstein: Personal Recollections.* Totowa, N.J.: Rowman and Littlefield, 1981.

Ricoeur, Paul. *The Conflict of Interpretations: Essays in Hermeneutics.* Translated by Willis Domingo, et al. Evanston, Ill.: Northwestern University Press, 1974.

———. *Hermeneutics and the Human Sciences.* Translated by John B. Thompson. New York: Cambridge University Press, 1981.

Rorty, Richard. *Contingency, Irony, and Solidarity.* Cambridge: Cambridge University Press, 1989.

———. *Philosophical Papers.* 2 vols. Cambridge: Cambridge University Press, 1991.

Rosenblum, Robert. *Modern Painting and the Northern Romantic Tradition: Friedrich to Rothko.* New York: Harper and Row, 1975.

Rothenberg, David. *Hand's End: Technology and the Limits of Nature.* Berkeley: University of California Press, 1993.

———. *Is It Painful to Think: Conversations with Arne Naess.* Minneapolis: University of Minnesota, 1990.

Sanford, Charles L. *The Quest for Paradise.* Urbana: University of Illinois Press, 1969.

Santmire, Paul H. "St. Augustine's Theology of the Biophysical World." *Dialog* 19 (1980): 19–45.

Schürmann, Reiner. *Heidegger on Being and Acting: From Principles to Anarchy.* Translated by Christine-Marie Gros. Bloomington: Indiana University Press, 1987.

Schwartz, Seymour I., and Ralph E. Ehrenberg. *The Mapping of America.* New York: Abrams, 1980.

Seamon, David. *A Geography of the Lifeworld: Movement, Rest, and Encounter.* London: Croom Helm, 1979.

———. "Heideggerian Thinking and Christopher Alexander's Pattern Language." Paper presented at the Annual Meetings of the Environmental Design Research Association, New York, June 1985.

———, ed. *Dwelling, Seeing, and Designing: Toward a Phenomenological Ecology.* Albany, N.Y.: SUNY Press, 1992.

Seamon, David, and Robert Mugerauer, eds. *Dwelling, Place and Environment.* Dordrecht, Holland: Nijhoff, 1985; reprint, New York: Columbia University Press, Morningside Editions, 1989.

Seamon, David, and Christine Nordin. "Marketplace as Place Ballet." *Landscape* 24 (1980): 35–41.

Sears, John F. *Sacred Places: American Tourist Attractions of the Nineteenth Century.* New York: Oxford University Press, 1989.

Seidel, George. *Martin Heidegger and the Pre-Socratics: An Introduction to His Thought.* Lincoln: University of Nebraska Press, 1964.

Shepard, Paul. *Man in the Landscape: A Historic View of the Esthetics of Nature.* New York: Knopf, 1967.

Smith, Henry Nash. *Virgin Land: The American Land as Symbol and Myth.* New York: Vintage, 1950.

Smith, H. Shelton, ed. *Horace Bushnell.* New York: Oxford University Press, 1965.

Soja, Edward. *Postmodern Geographies: The Reassertion of Space in Critical Social Theory.* New York: Verso, 1989.

Spring, David, and Ellen Spring, eds. *Ecology and Religion in History.* New York: Harper and Row, 1974.

Standing Bear, Luther. *Land of the Spotted Eagle.* Lincoln: University of Nebraska Press, 1978.

Starr, Kevin. *Americans and the California Dream 1850–1915.* Santa Barbara, Cal.: Peregrine Smith, 1973.

———. *Inventing the Dream: California through the Progressive Era.* New York: Oxford University Press, 1985.

Staten, Henry. *Wittgenstein and Derrida.* Lincoln: University of Nebraska Press, 1984.

Steck, O. H. *World and Environment.* Biblical Encounters Series. Nashville: Abingdon, 1980.

Stilgoe, John R. *Common Landscape of America: 1580 to 1845.* New Haven, Conn.: Yale University Press, 1982.

Sutton, S. B., ed. *Civilizing American Cities.* Cambridge, Mass.: MIT Press, 1971.

Sweeney, J. Gray. "The Advantages of Genius and Virtue: Thomas Cole's Influence, 1848–58." In *Thomas Cole: Landscape into History,* edited by William H. Truettner and Alan Wallach, 113–135. Washington, D.C.: National Museum of American Art, Smithsonian Institution, 1994.

———. *The Columbus of the Woods: Daniel Boone and the Typology of Manifest Destiny.* St. Louis: Washington University Gallery of Art, 1992.

———. "'Endowed with Rare Genius': Frederic Edwin Church's *To the Memory of Cole.*" *Smithsonian Studies in American Art* 2, no. 1 (Winter 1988): 45–72.

———. *Natural Divinity.* Unpublished research project, Smithsonian Institution Senior Fellowship, 1984–1985.

———. "The Nude of Landscape Painting: Emblematic Personification in the Art of the Hudson River School." *Smithsonian Studies in American Art* 3, no. 4 (Fall 1989): 43–65.

———. *Themes in American Painting.* Grand Rapids, Mich.: The Grand Rapids Art Museum, 1977.

Taylor, Mark. "Architecture of Pyramids." *Assemblage* 5 (Feb. 1988): 17–27.

———. "Deadlines: Approaching an Architecture." *Threshold* 4 (Spring 1988): 20–27.

Thomas, Keith. *Man and the Natural World: History of the Modern Sensibility.* New York: Pantheon, 1983.

Truettner, William H., ed. *The West as America: Reinterpreting Images of the Frontier, 1820–1920.* Washington, D.C.: Smithsonian Institution Press, 1991.

Truettner, William H., and Alan Wallach, eds. *Thomas Cole: Landscape into History.* Washington, D.C.: National Museum of American Art, Smithsonian Institution, 1994.

Tuckerman, Henry T. *Book of the Artists: American Artist Life Comprising Biographical and Critical Sketches of American Artists.* New York: James E. Carr, 1967 (1867).

Vance, James. "California and the Search for the Ideal." *Annals of the Association of American Geographers* 62 (Jan. 1972): 185–210.

Van Every, Dale. *The Disinherited: The Lost Birthright of the American Indian.* New York: Avalon, 1966.

Venturi, Robert. *Complexity and Contradiction in Architecture.* The Museum of Modern Art Papers on Architecture, no. 1. New York: Doubleday, 1966.

Virilio, Paul. *Lost Dimension.* New York: Semiotext(e), 1991.

Von Rad, Gerhard. *Genesis: A Commentary.* Philadelphia: Westminster, 1956.

von Wright, G. H. "Biographical Sketch." In *Ludwig Wittgenstein,* edited by Norman Malcolm, 1–22. New York: Oxford University Press, 1958.

Wallach, Alan. "Thomas Cole: Landscape and the Course of Empire." In *Thomas Cole: Landscape into History,* edited by William H. Truettner and Alan Wallach, 23–112. Washington, D.C.: National Museum of American Art, Smithsonian Institution, 1994.

Weinsheimer, Joel C. *Gadamer's Hermeneutics: A Reading of Truth and Method.* New Haven, Conn.: Yale University Press, 1985.

White, Lynn. "The Historical Roots of Our Ecological Crisis." In *Ecology and Religion in History,* edited by David Spring and Ellen Spring, 15–31. New York: Harper and Row, 1974.

Wijdeveld, Paul. *Ludwig Wittgenstein, Architect.* Cambridge: MIT Press, 1994.

Wilkins, Thurman. *Thomas Moran: Artist of the Mountains.* Norman: University of Oklahoma Press, 1966.

Williams, George H. *Wilderness and Paradise in Christian Thought: From the Garden of Eden and the Sinai Desert to the American Frontier.* New York: Harper and Row, 1962.

Wilmerding, John. *American Light: The Luminist Movement.* Washington, D.C.: National Gallery of Art, 1986.

Wilson, Alexander. *The Culture of Nature: North American Landscape from Disney to the Exxon Valdez.* Cambridge, England: Blackwell, 1992.

Winters, Yvor. *In Defense of Reason.* Denver: Alan Swallow, 1947.

———. *The Function of Criticism.* Denver: Alan Swallow, 1957.

Wittgenstein, Hermine. "My Brother Ludwig." Translated by Bernhard Leitner. In *Ludwig Wittgenstein: Personal Recollections,* edited by Rush Rees, 6. Totowa, N.J.: Rowman and Littlefield, 1981.

Wittgenstein, Ludwig. *Culture and Value.* Edited by C. H. von Wright and translated by Peter Winch. Oxford, England: Blackwell, 1980.

———. *Philosophical Investigations.* Translated by G. E. M. Anscomb. New York: Macmillan, 1965.

———. *Tractatus Logico-Philosophicus.* Translated by D. F. Pears and B. F. McGuinness. London: Routledge and Kegan Paul, 1963.

Wyatt, David. *The Fall into Eden: Landscape and Imagination in California.* Cambridge: Cambridge University Press, 1986.

Yuha, Chunksa, and James E. Ricketson. "Glossary of Lakotah Words." Appendix to Ruth Beebe Hill, *Hanta Yo,* 1093–1109. New York: Warner, 1979.

Zukin, Sharon. *Landscapes of Power: From Detroit to Disney World.* Berkeley: University of California Press, 1991.

Index